Taguchi Techniques for Quality Engineering

Loss Function, Orthogonal Experiments, Parameter and Tolerance Design

Phillip J. Ross

Second Edition

McGraw-Hill

New York   San Francisco   Washington, D.C.   Auckland   Bogotá
Caracas   Lisbon   London   Madrid   Mexico City   Milan
Montreal   New Delhi   San Juan   Singapore
Sydney   Tokyo   Toronto

Library of Congress Cataloging-in-Publication Data

Ross, Phillip J.
    Taguchi techniques for quality engineering : loss function,
orthogonal experiments, parameter and tolerance design / Phillip J.
Ross.—2nd ed.
            p.   cm.
        Includes index.
        ISBN 0-07-053958-8
        1. Quality control—Statistical methods.   2. Taguchi methods
(Quality control)   3. Engineering design—Statistical methods.
    I. Title.
    TS156.R67   1996
    658.5′62′015195—dc20   95-15415
                                                                CIP

1 2 3 4 5 6 7 8 9 0   DOC/DOC   9 0 0 9 8 7 6 5

ISBN 0-07-053958-8

*The sponsoring editor for this book was Harold B. Crawford, the editing
supervisor was Jane Palmieri, and the production supervisor was
Suzanne W. B. Rapcavage. It was set in Century Schoolbook by Estelita
F. Green of McGraw-Hill's Professional Book Group composition unit.*

Printed and bound by R. R. Donnelley & Sons Company.

McGraw-Hill books are available at special quantity discounts to use
as premiums and sales promotions, or for use in corporate training pro-
grams. For more information, please write to the Director of Special
Sales, McGraw-Hill, 11 West 19th Street, New York, NY 10011. Or contact your local bookstore.

 This book is printed on recycled, acid-free paper containing
a minimum of 50% recycled, de-inked fiber.

To Emily,
who is always first in my book

A chasm.
A bridge to build.
Not as a monument in itself,
but to make it easier for others to follow.

Contents

Preface

Taguchi Techniques for Quality Engineering is intended as a guide and reference source for industrial practitioners (managers, engineers, and scientists) involved in product or process experimentation and development. Most engineers are familiar with setting up tests to model actual field conditions and the cause-effect relationship of design to performance; however, their knowledge of a proper testing strategy is usually limited. When engineers have had exposure to experimental design, typically, the reaction is to deem the approach too costly and time consuming because of full-factorial designs.

It is most unfortunate that people are not aware of the potential savings in test time and money offered by more efficient testing strategies. Not only are savings in test time and cost available but also a more fully developed product or process will emerge with the use of better experimental strategies. The second edition of this book has been reorganized to emphasize the *design of experiments (DOE) process* which will enable people to easily understand how to utilize this quality improvement method. Also, example experiments are used throughout Chaps. 2 to 6 to demonstrate the application of the DOE process.

The Taguchi philosophy provides two tenets: (1) the reduction in variation (improved quality) of a product or process represents a lower loss to society, and (2) the proper development strategy can intentionally reduce variation. Again, most managers and engineers are not aware of the economics of improved quality and the techniques to achieve higher quality at lower costs.

This book addresses the basic testing and development strategies that have allowed some Japanese companies to successfully become world economic competitors. Many Japanese engineers since the mid-1960s have had Taguchi training. In 1985, Nippon Denso, a Toyota affiliate, ran over 2500 experiments concerning automotive electrical products, for example.

By using and understanding the Taguchi methods, managers and engineers will realize what is required to put western product development and quality back into the competition. Today's managers and engineers must have a certain amount of exposure to these methods before they can appreciate how much improvement in testing and development strategies can be made. The text takes a user-oriented, hands-on approach for working engineers or scientists and their immediate management to develop initial expertise in the Taguchi methodology. Too often, texts on the subject of design of experiments focus on the analytical phase of the process, but most novices wonder how the experiment was initially developed. This book should be quite useful to the statistically inexperienced engineer who would have some difficulty understanding and utilizing a traditional text concerning designed experiments.

It is hoped that *Taguchi Techniques for Quality Engineering* will bridge the gap for the industrial user, eventually making American companies more competitive with their products in a world market. World-class quality will be a requirement for corporations to remain lucrative, let alone highly competitive, as more and more companies embrace the Taguchi methods.

Acknowledgments

My indebtedness is to Bill Diamond who introduced me to experimentation based on orthogonal arrays, the Hadamard matrices. I know he would hate to admit it, but the Taguchi matrices are mathematically and statistically equivalent to Hadamard's. The approaches used by the two experimenters may be different but the matrices are not.

I would like to give my thanks to Kathy Layne and Steve Abney, two very dear friends, who acted as statistical consultants, an editorial staff, and sympathetic ears during the creation of the first edition. My thanks also to those who suffered through some of the pilot classes and seminars on designed experiments; your contribution has been the engineer's point of view and, for me, a better appreciation of what the customer expects from this product. In fact, it is this customer's point of view that precipitated the second edition and the reorganization to make design of experiments even more user friendly. And lastly, thanks to my family for the tolerance of all the lost evenings and weekends as Dad lingered over the PC.

The author expresses gratitude to the American Supplier Institute, Inc., Center for Taguchi Methods, for granting permission to reproduce the orthogonal array, triangular table, and related linear graphs. They are contained in ASI's edition of *Orthogonal Arrays and Linear Graphs* (1987).

Phillip J. Ross

Introduction

Taguchi addresses quality in two main areas: off-line and on-line quality control (QC). Both of these areas are very cost sensitive in the decisions that are made with respect to the activities in each. Off-line QC refers to the improvement of quality in the product and process development stages. On-line QC refers to the monitoring of current manufacturing processes to verify the quality levels produced. The off-line portion of QC is addressed in this text because of the paucity of materials on this phase of Taguchi methods and the positive impact on cost that is obtained by improving quality at the earliest times in a product life cycle.

The on-line phase is covered by many texts as a dimensional approach to quality control, typified by statistical process control, or SPC, and for this reason will not be addressed in this text. The Taguchi on-line QC approach is a cost quality control perspective and someday should be recognized as an alternative quality control system.

This text reviews the basic aspects of off-line QC developed by Taguchi. There are several more sophisticated concepts for off-line QC that are not covered in this text but should be pursued by the experimenter after initial work with these methods. Also, education and training in general statistical methods is recommended. In particular, other designed experiment texts will be valuable for the experimenter as background.

Of particular emphasis in the second edition is the design of experiments process. The second edition has been substantially revised to introduce topics in order of their application to completing one designed experiment. A short description of the chapter contents follows.

Chapter 1. The Economics of Reducing Variation

The economics (cost reduction) of reducing variation is a subject that is not addressed widely at this time. Taguchi uses a different cost model for product characteristics than is typically used, which places more emphasis on reducing variation, particularly when the total product variation is within the specification limits for the product. This chapter covers the conventional viewpoint of cost versus specification limits and introduces the Taguchi model for cost versus specification limits: the loss function. The Taguchi methodology ascribes to the approach that the lowest loss to society represents the product with the highest quality. Higher product quality by definition means less variation of a product characteristic. The difference between conventional and Taguchi approaches to higher quality is that one proposes that higher quality costs more and the other proposes that higher quality costs less. The loss function is a mathematical way of quantifying the cost as a function of product variation, which answers the question of whether further reduction of variation will reduce costs.

Also discussed is the difference between off-line and on-line QC and where the responsibility falls for that portion of total quality control. The different types of loss functions are described for the typical types of product characteristics, such as higher is better, nominal is best, and lower is better.

Chapter 2.   The Design of Experiments Process

The design of an experiment (DOE) is not a simple one-step process but is actually a series of steps which must follow a certain sequence for the experiment to yield an improved understanding of product or process performance. The DOE process is made up of three main phases: the planning phase, the conducting phase, and the analysis/interpretation phase. The steps in the DOE process are generically the same regardless of the experiment design, which is chosen to evaluate factorial effects. The experiment design can be anywhere between a full-factorial experiment and a very small fractional-factorial experiment. Many texts on the subject of designed experiments emphasize the analytical phase of the DOE process; however, positive experimental results are dependent upon the planning of the experiment and not on the analysis. This chapter provides a guideline for generating successful experimental results. This portion of the text also introduces the initial steps in the planning phase of experimentation.

Chapter 3.   Orthogonal Array Selection and Utilization

A major step in the DOE process is the determination of the combination of factors and levels which will provide the experimenter with the desired information. One approach is to utilize a fractional-factorial approach whenever there are several factors involved, and this may be accomplished with the aid of orthogonal arrays. Orthogonal arrays are introduced from the viewpoint of the pragmatist who is always trying to make product or process improvement decisions with the minimum amount of test data. Using a minimal amount of test data is not necessarily a problem in itself; however, the considerations of what may make up a valid experiment from a risk viewpoint are seldom considered by the typical experimenter. The statistical aspects of the size of an experiment are discussed in conjunction with the amount of information required to be evaluated during the experiment. The aspects of designing a simple orthogonal experiment are discussed, including the handling of various factors and interaction effects. More complex situations, such as multiple-level factors and factors with different numbers of levels, which require special orthogonal array adaptations, are also covered. A typical situation that confronts an engineer is handled with nested experiments. Many times, discrete variables such as different materials cause a portion of the experiment to be unable to be exposed to the variation of other factors. This situation requires the use of a nested experiment to allow easy analysis of the data. Engineers will find themselves running into this fairly frequently. Two methods of handling a mixture of factors with different numbers of levels are covered and a component identification technique is described for problems where retest is possible. Many nondestructive tests could benefit from this test strategy.

Chapter 4. Conducting Tests

The logistics of collecting experimental data are addressed in this part of the text. Two statistical aspects of conducting tests, sample size and randomization, are addressed. The sample size affects the sensitivity of the experiment, and randomization protects the experimenter from unknown influences which may bias the experimental results and subsequent decisions. Different randomization strategies are applied to typical product, process, and production situations.

The experimenter is provided with some guidelines to determine whether the experimental results will provide positive information concerning the factors investigated. Also, emphasis is placed on utilizing variable data whenever possible.

Chapter 5. Analysis and Interpretation
Methods for Experiments

This chapter introduces several methods for analyzing and interpreting experimental results. The orthogonal array structure provides the opportunity to utilize some very simple methods, such as the observation method and column effects method. More sophisticated methods are covered, starting with the basics of conducting an analysis of variance for a particular set of data and how the ANOVA techniques work from a statistical basis. The examples are extremely simple and easy to follow because the purpose of the book is to teach a graduate-level person the mechanics of ANOVA, not to practice basic arithmetic. The F test, a basic statistical test of comparisons of products, is introduced and explained in a simple manner as a tool to make decisions after an ANOVA is completed. Analysis of variance is applied to the orthogonal array type of designed experiment. Polynomial decomposition of variance is introduced with a discussion of what information is gained when running a multiple-level versus a two-level experiment. All of the problems discussed up to this point concern variable data, such as weight, diameter, and voltage, which are on a continuous scale from high to low. However, some experiments do not lend themselves to this type of measurement system. Results are on a discontinuous scale, such as good or bad, passing or failing, meets specification or doesn't meet specification. Such ratings become attribute characteristics and require a different type of analysis than does variable data. The factors may be allocated to an experiment in the same manner, but the analysis can be modified for attribute data.

Chapter 6.   Confirmation Experiment

The purpose of a confirmation experiment, especially when low-resolution, small fractional-factorial experiments are used, is emphasized. Methods of estimating different values such as the mean, confidence intervals, and capability from the experimental data are discussed. Two methods of estimating the mean are used: standard and omega transformation. These estimates are used as predictions of the results of a confirmation experiment which validates the conclusions drawn from the previous round of experimentation. A flowchart is provided as an aid to making decisions about further tests after the first confirmation experiment is conducted.

Chapter 7.   Parameter Design

The main thrust of Taguchi methods is the use of parameter design, which is the ability to design a product or process to be resistant to various environmental factors that change continuously with customer use. To determine the best design of a product or process requires the use of a strategically designed experiment which exposes the product or process to the varying environmental conditions.

Taguchi refers to these variations in customer use as *noise factors*. The analysis of the experimental results uses a signal-to-noise ratio to aid in the determination of the best product or process designs. Nonlinear response characteristics of products or processes can be used to the engineer's advantage if the proper design philosophy is employed. The concept of dynamic characteristics is introduced, as well, which requires a different orthogonal array arrangement for experimentation. Parameter design is to achieve high quality at relatively low cost in the product or process by selecting the appropriate specifications.

Chapter 8.   Tolerance Design

Tolerance design is to achieve high quality at some cost in the product or process by tightening tolerances on specifications. This approach is used when parameter design has not achieved the quality level desired by the customer. Tolerance design utilizes information from the ANOVA of the experiment to determine which specifications are most appropriate to tighten.

The Economics of Reducing Variation

1-1 The Meaning of Quality

Products have characteristics that describe their performance relative to customer requirements or expectations. Characteristics such as fuel economy of a car, the weight of a package of breakfast cereal, the power losses of a home hot water heater, or the breaking strength of fishing line are all examples of product characteristics that are of concern to customers at one time or another.

The quality of a product is measured in terms of these characteristics. Quality is related to the loss to society caused by a product during its life cycle. A truly high quality product will have a minimal loss to society as it goes through this life cycle. The loss a customer sustains can take many forms, but it is generally a loss of product function or properties. Other losses are time, pollution, noise, etc. If a product does not perform as expected, the customer senses some loss. After a product is shipped, a decision point is reached; it is the point at which the producer can do nothing more to the product. Before shipment, the producer can use expensive or inexpensive materials, use an expensive or inexpensive process, etc.; but once shipped, the commitment is made for a certain product expense during the remainder of its life.

Quality has but one true evaluator: the customer. A "quality circle" that describes this situation is shown in Fig. 1-1. The customer is judge, jury, and executioner in this model. Customers vote with their wallets on which products meet their requirements, including price and performance. The birth of a product, if you will, is when a designer takes information from the customer (market) to define what the

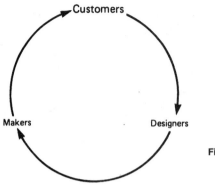

Figure 1-1   Quality circle.

customer wants, needs, and expects from a particular product. Sometimes, a new idea (high technology) creates its own market, but once a competitor can duplicate the product, the technological advantage is lost.

The designer must take the customer's wants, needs, and expectations and translate them into product specifications, which include drawings, dimensions, tolerances, materials, processes, tooling, and gaging. The makers use this information, along with the prescribed machinery, to fabricate the product. The product is then delivered via marketing channels to the customer. To satisfy the customer, the product must arrive in the right quantities, at the right time, at the right place, and provide the right functions for the right period of time. All of this must be available to the customer at the right price, too. This is a tough order to fill, but the simplest definition of high quality is a happy customer. Customers should become more endeared to a product the more it is used. Customer feedback to the designers and makers comes in terms of the number of products sold and the warranty, repair, and complaint rate. Increasing sales volume and market share with low warranty, repair, or complaint rates translates to happy customers.

1-2 Goalpost Philosophy

Today in America it is quite popular to take a very strict view of what constitutes quality. In his book, Crosby supports the position that a product made according to the print, within permitted tolerance, is of high quality.* This strict viewpoint embraces only the designers and

---

*Philip B. Crosby, *Quality Is Free,* McGraw-Hill, New York, 1979.

the makers. This is the "goalpost" syndrome.[†] What is missing from this philosophy is the customer's requirements. A product may meet print specifications, but if the print does not meet customer requirements, then true quality cannot be present. For example, customers buy TVs with the best picture, not ones that necessarily meet specifications.

Another example showing that the goalpost syndrome contradicts the customer's desires is as follows. Batteries supply a voltage to a light bulb in a flashlight. There is some nominal voltage, let us say 3 volts, that will provide the brightest light but will not burn out the bulb prematurely. Customers want the voltage to be as close to the nominal voltage as possible, but battery manufacturers may be using a wider tolerance than allowed by the battery specification. As a result, some flashlights burn dimly and others burn brightly but burn out the bulbs prematurely. Customers want the product close to nominal all the time, and producers want to allow the product to vary to the limit of the specifications; how can these seemingly incongruent ideas be brought into harmony?

1-3   Taguchi Loss Function

The Taguchi loss function recognizes the customer's desire to have products that are more consistent, part to part, and a producer's desire to make a low-cost product. The loss to society is composed of the costs incurred in the production process as well as the costs encountered during use by the customer (repair, lost business, etc.). To minimize the loss to society is the strategy that will encourage uniform products and reduce costs at the point of production and at the point of consumption. Let's look at a comparison of the goalpost and loss function philosophies with an example.

1-4   Comparison of Philosophies

When the hood of a typical automobile is opened, a mechanism may be in place which automatically holds the hood in the open position. The force required to close the hood from this position is important to the customer. If the amount of force required is too high, then a weaker individual may have difficulty in closing the hood and ask for the

---

[†]In football, a team is awarded three points for a field goal regardless of where the ball passes through the uprights, whether exactly midway between the uprights or far to left or right. There is no additional reward for being in the middle.

mechanism to be adjusted. If the amount of force required is too low, then the hood may come down when a gust of wind hits it, and again the customer will ask for it to be adjusted. The engineering specifications and detail and assembly drawings call out a particular range of force values for the hood assembly. A range must be used, since all hoods cannot be exactly the same; a lower limit (LL) and an upper limit (UL) are specified. If the force is a little high or low, the customer may be somewhat dissatisfied but may not ask for an adjustment to the hood. A goalpost view of this situation is shown graphically in Fig. 1-2.

The goalpost philosophy says that as long as the amount of closing force is within the zone shown as the customer's tolerance, this would be satisfactory—no problem. If the amount of closing force is smaller than the lower limit or greater than the upper limit of the customer's tolerance, then the hood would have to be adjusted at some expense, say $50, to be borne by the manufacturer (warranty).

As a customer, the closer the amount of closing force is to the nominal, or target, value, the happier you are. If the amount of force is a little low or a little high, you sense some loss. If the amount of force is even lower or higher, you would sense a greater loss; the hood would come down more frequently or be uncomfortably hard to close. When the amount of force reached the customer tolerance limits, the typical customer would complain about the hood. But what is the real difference between a closing force indicated by points A and B on the goalpost graph? It appears from a producer's viewpoint that the difference is the total cost of adjustment. From a customer's viewpoint there is

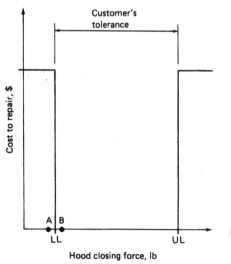

Figure 1-2 Goalpost syndrome.

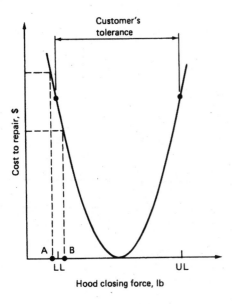

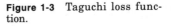

Figure 1-3   Taguchi loss function.

very little difference in a hood that falls down just a little bit more easily. A better model for the cost versus closing force is shown in Fig. 1-3.

This curve, the *loss function,* more nearly describes the real situation. If the amount of closing force is near the nominal value, there is no cost or very low cost associated with the hood. The farther the force gets from the nominal force, the greater the cost associated with that force is, until the customer's limit is reached at which the cost equals the adjustment cost. This model quantifies the slight difference in cost associated with a hood closing force of force A and force B. This is a fictitious example, but let's look at an actual case from Japan.

1-5   Case Study: Polyethylene Film*

A supplier in Japan made a polyethylene film with a nominal thickness of 0.039 in (1.0 mm) that is used for greenhouse coverings. The customers want the film to be thick enough to resist wind damage but not too thick to prevent passage of light. The producers want the film to be thinner to be able to produce more square feet of material at the same cost. A plot of these contradictory desires is shown in Fig. 1-4. At the time the national specifications for film thickness stated that the film should be 0.039 in ±0.008 in (0.2 mm). A company that made

---

*Genichi Taguchi and Yu-in Wu, *Off-Line Quality Control,* Central Japan Quality Control Association, Nagaya, 1979, pp. 7–8.

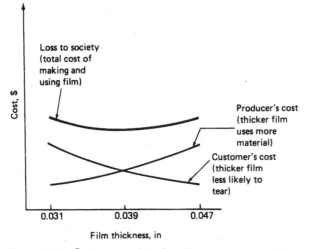

Figure 1-4 Costs associated with greenhouse film. (*Reproduced from Genichi Taguchi and Yu-in Wu,* Off-Line Quality Control, *Central Japan Quality Control Association, Nagaya, 1979, p. 7. Used by permission.*)

this film could control film thickness to ±0.0008 in (0.02 mm) consistently. The company made an economic decision to reduce the nominal thickness to 0.032 in (0.82 mm), and, with their ability to produce film within 0.0008 in of the nominal, all of the product would meet the national specification. This would reduce manufacturing costs and increase profits.

However, that same year strong typhoon winds caused a large number of the greenhouses to be destroyed. The cost to replace the film had to be paid by the customer, and these costs were much higher than expected. Both of these cost situations can be seen in Fig. 1-4. What had not been considered by the film producer was the fact that the customer's cost would rise while the producer's cost was falling. The loss function, loss to society, is the upper curve, which is the sum of the producer's and customer's curves. This curve does show the proper thickness for the film to minimize loss to society, and this is where the nominal value of 0.039 in is located.

Looking at the loss function, one can easily see that as the film gets thicker from the nominal of 0.039 in, the producer is losing money, and when the film gets thinner from the nominal, the customer is losing money. The producer is obligated by being part of society to fabricate film with a nominal of 0.039 in and to reduce variation of that thickness to a low amount. In addition, it will save money for society to further reduce the manufacturer's capability from ±0.0008 in (losses are lower closer to the nominal value).

If the producer does not attempt to hold the nominal thickness at 0.039 in and causes additional loss to society, then this is worse than stealing from the customer. If someone steals $10, the net loss to society is zero; someone has a $10 loss and the thief has a $10 gain. If, however, the producer causes an additional loss to society, everyone in society has suffered some loss. A producer who saves less money than the customer spends on repairs has done something worse than stealing from the customer. After this experience, the national specification was changed to make the average thickness produced 0.039 in. The tolerance was left unchanged at ±0.008 in.

1-6   Japan's Desire for Low Loss

One must understand Japan's position in worldwide economic competition to see how this philosophy was established. Japan is an island with limited natural resources except for a large group of people. The only method by which Japan can survive economically is to import materials, add value to the materials by processing, and export products. Much of Japan's economic success has been based on high-value-added products. High technology products, such as computers, are a natural for this approach. Success in the business of converting resources from one state (raw) to another state (finished) is due in part to the efficiency of this process. Efficiency of adding value to materials is equivalent to a low loss in the process. Low loss to society fits neatly within the niche of competing effectively in a value-added product arena.

1-7   Case Study: Tool Wear in a Process

This case study is intended to show a cost-oriented approach to quality control. With a very capable process, the goalpost philosophy allows tool wear to produce parts which vary from one specification limit to the other. The loss function economically justifies a different quality control approach. To illustrate this comparison, a new concept must first be introduced: a frequency, or probability of occurrence, distribution of products according to a measured value of a performance characteristic.

A typical frequency distribution for the heights of NBA basketball players might look like the histogram in Fig. 1-5. This graph says the height (performance characteristic of interest) of 80 in (2.03 m) is much more likely to occur than any other height, and this likelihood falls off as height goes higher or lower. This shape of distribution is known as a normal distribution, and is typically illustrated in the manner shown. Frequency distributions do not have to be bell-shaped

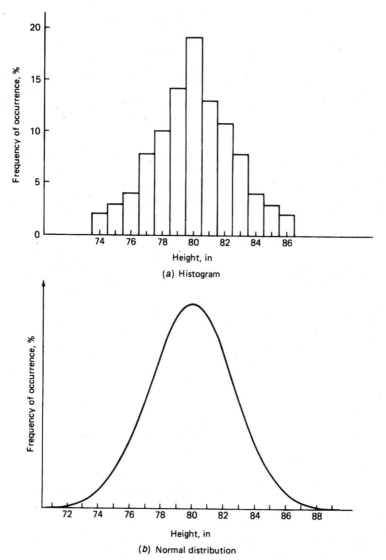

Figure 1-5 Frequency distributions of heights of NBA basketball players. (*a*) Histogram, (*b*) normal distribution.

as is this normal distribution, but many manufacturing processes and many natural phenomena produce results in this manner.

An important aspect of product and process quality is the relative widths of the distribution and the specification limits of the product. If the distribution is narrower than the specification limits, then it is possible to make all or nearly all of the parts to match the printed specification. This is especially true if the distribution is centered within the

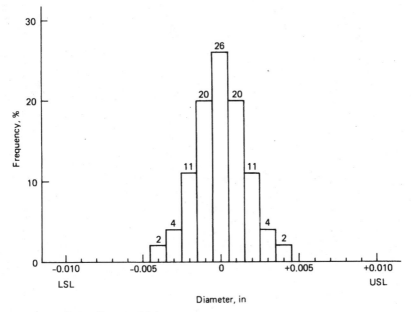

Figure 1-6 Outer diameter histogram.

specification limits. All of this is very desirable from the goalpost point of view; however, what about the loss function viewpoint? An example will demonstrate the application of the loss function.

A typical tolerance for a machined part—let's say an outer diameter—might be ±0.010 in (±0.25 mm). A process that would be very capable with respect to this tolerance would be modeled by the group of parts shown in Fig. 1-6. This histogram is for 100 separate parts.

The Taguchi loss function quantifies the variability present in a process. If a part reaches the end of the manufacturing line with a diameter exceeding the upper or lower limit, the part should be scrapped at a cost assumed to be $4.00. The scrap cost is only one aspect of loss to society. Presumably, the specifications are related to the reliability of the product; as the specification limits are approached, the product is less likely to provide satisfaction to the customer. If the product fails to perform satisfactorily, then other losses are incurred by the manufacturer or the customer, which makes scrap loss a conservative (low) estimate of loss to society.

Using this cost as a reference value, a loss function can be constructed for this situation, as shown in Fig. 1-7. Taguchi uses the mathematical equation to model this picture of cost versus outer diameter:

$$L = k(y - m)^2 \qquad (1\text{-}1)$$

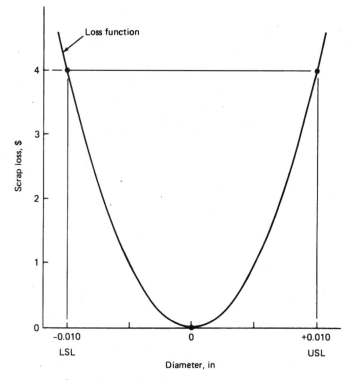

Figure 1-7 Outer diameter loss function.

In this equation, L is the loss associated with a particular diameter value y, m is the nominal value of the specification, and the value of k is a constant depending on the cost at the specification limits and the width of the specification. In this example

$$L = k(y - m)^2$$
$$\$4.00 = k(\text{LSL} - m)^2$$

The lower specification limit (LSL) is substituted into the equation, which is where the \$4.00 loss is incurred. The upper specification limit also could be used for this calculation. Solving for k,

$$k = \frac{\$4.00}{(\text{LSL} - m)^2}$$

Given that $m = 0.0$ in (nominal value),

$$k = \frac{\$4.00}{(-0.010 - 0.0)^2}$$
$$k = \$40{,}000 \text{ per in}^2$$

Therefore,

$$L = 40,000(y - 0.0)^2 \qquad (1\text{-}2)$$

Now the loss associated with any part can be computed depending on the value of its diameter. For instance, a part with a diameter of + 0.003 in (0.08 mm) costs

$$L = 40,000(0.003 - 0.0)^2 = \$.36$$

This is the loss per unit for each part shipped with an outer diameter of +0.003 in.

The average cost per part for a particular group of parts also can be determined from this loss function. This can be accomplished in two ways. One way is to use the histogram of outer diameters and calculate the total cost for each value in the histogram, add the costs for each value, and divide by the total number of parts in the histogram. Using the dimensions for the parts shown in Fig. 1-6, one can calculate the loss. The numbers at the top of each bar indicate the quantity of parts with a particular outer diameter. By using the loss function, Eq. 1-2, the cost of a part of a particular diameter can be calculated.

$$L(-0.002) = \$40,000(-0.002 - 0.0)^2 = \$.16$$

Since there are 11 parts having this value, the loss for all the parts having an outer diameter of − 0.002 in is \$1.76. If we calculate this cost for each diameter, then the results are as shown in Table 1-1. The average loss per part is then

$$L = \frac{\$10.56}{100 \text{ parts}} = \$0.11 \text{ per part}$$

TABLE 1-1   Cost of Parts versus Outer Diameter

Diameter	Loss/part	Total parts	Total loss
−.005	\$1.00	0	\$ 0.00
−.004	.64	2	1.28
−.003	.36	4	1.44
−.002	.16	11	1.76
−.001	.04	20	.80
.000	.00	26	0.00
.001	.04	20	.80
.002	.16	11	1.76
.003	.36	4	1.44
.004	.64	2	1.28
.005	1.00	0	0.00
GRAND TOTALS		100	\$10.56

A second method of estimating average loss per part entails using the loss equation in a slightly modified form. Mathematically, this calculation is equivalent to using the average value of the $(y - m)^2$ portion of the loss equation. Expanded, this is

$$L = \frac{k[(y_1 - m)^2 + (y_2 - m)^2 + \ldots + (y_N - m)^2]}{N} \tag{1-3}$$

where N = number of parts sampled

If all the $(y - m)$ values are squared, added together, and divided by the number of items, then the result is the desired value. For a large number of parts, the average loss per part is equivalent to

$$L = k[S^2 + (\bar{y} - m)^2] \tag{1-4}$$

S^2 = variance around the average, $\bar{y}$
$\bar{y}$ = average value of y for the group
$(\bar{y} - m)$ = offset of the group average from the nominal value m

Mathematically, Eqs. 1-3 and 1-4 are equivalent. A proof of the preceding situation is located in App. E.

For the preceding group of parts, the values of S^2 and $\bar{y}$ can be calculated using any statistical calculator and are

$$S^2 = 2.64 \times 10^{-6} \text{ in}^2 \, (1.72 \times 10^{-3} \text{ mm}^2)$$

$$\bar{y} = 0.0$$

$$L = k[S^2 + (\bar{y} - m)^2]$$

$$L = 40{,}000[2.64 \times 10^{-6} + (0.0 - 0.0)^2]$$

$$L_1 = \$.11 \text{ per part}$$

This second method uses the general loss function for a nominal-is-best situation.

This example demonstrates the loss associated with a distribution that has a very low process capability ratio and is centrally located on the target value of 0.0 in. What if this machined feature causes the machining tool to wear, which subsequently causes an increasing diameter on sequential parts? Traditional quality control thinking (goalpost) would allow the situation shown in Fig. 1-8. The machine would be adjusted to the low side and allowed to wear until the high side was reached. Again, very few, if any, parts are outside the specification limits. What is the loss associated with each part on this basis?

The loss function uses the average and variance (standard deviation squared) for a group of parts to calculate loss. For a group of parts where the distribution average is continually increasing, the variance is

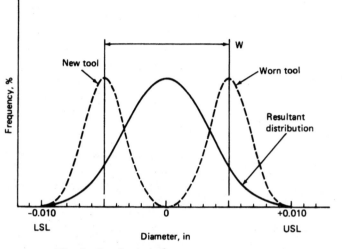

Figure 1-8 Distribution due to tool wear.

$$S^2_{\text{Resultant}} = S^2_{\text{Original}} + \left(\frac{W^2}{12}\right)$$

where W is the width of tool wear allowed. In this example,

$$S^2_O = 2.64 \times 10^{-6} \text{ in}^2$$
$$W = 0.010 \text{ in}$$

$$S^2_R = 2.64 \times 10^{-6} + \left(\frac{0.010^2}{12}\right) = 1.10 \times 10^{-5}$$

The loss per part is then

$$L = k[S^2 + (\bar{y} - m)^2]$$
$$L_2 = \$40,000[1.10 \times 10^{-5} + (0 - 0)^2] = \$0.44$$

One can readily see that the loss of a centered distribution is less than the loss of a distribution which traverses the specification range.

$$L_1 < L_2$$

But to obtain the low loss as in the first situation, the machine would have to be adjusted after each part. For the second situation, the machine would not have to be adjusted nearly as often. There must, therefore, be some economic compromise between the loss function and adjustment costs.

The additional loss caused by situation 2 is

$$L_{\text{Additional}} = L_2 - L_1$$

$$L_A = k\left[S^2 + \left(\frac{W^2}{12}\right)\right] - k(S^2)$$

$$L_A = k\left(\frac{W^2}{12}\right)$$

The additional loss is due only to allowing the process to traverse across the tolerance band. The larger W becomes, the larger the loss. However, the larger W becomes, the less frequently the machine must be adjusted and the lower the adjustment costs. A plot of these two cost functions is shown in Fig. 1-9. The lowest total loss (optimum) is near the intersection of the two curves. The minimum value of $[L_A+(\text{adjustment cost per part})]$ is located at the optimum adjustment width W.

C_A = adjustment cost per part for N parts
C_A = total machine adjustment cost divided by N
N = number of parts made since last adjustment
$N = W/R$
R = wear rate of tool (wear per part)

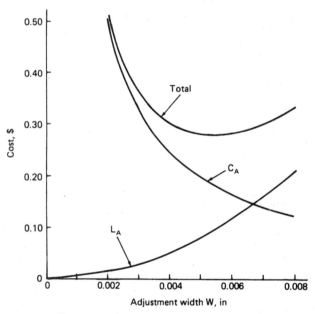

Figure 1-9 Cost comparison.

Substituting,

$$C_A = R \times \left(\frac{\text{total machine adjustment cost}}{W} \right)$$

Table 1-2 shows the optimum adjustment width for this example if these values are assumed

$k = 40,000$
$R = 0.0005$ in per part
$\$2.00 = $ total machine adjustment cost

The true optimum loss is at 0.005 in total wear span, but the additional loss of going up to 0.006 in total span is very minimal (\$0.003), which is beyond the accuracy of the loss function. Therefore, with a tool wear rate of 0.0005 in per part and an adjustment width of 0.006 in, the machine should be adjusted after 12 parts are produced. The average value for the outer diameter should be −0.003 in after adjustment, and the tool should be allowed to wear until the average value for the outer diameter becomes +0.003 in, providing a tool wear span of 0.006 in centered on the nominal value of 0.0 in.

Recall that the optimum adjustment interval depended on the scrap loss of one part, which was a conservative estimate of societal loss for the specification limit. If a larger loss is attributed to a part at the specification limit, the k value is increased, the loss function cost is increased, and the adjustment interval is subsequently decreased. When greater losses are incurred, greater uniformity of the product is necessitated.

Another contradiction also exists between the goalpost and loss function approach in this situation. If the process becomes more capable than depicted here, the goalpost approach would allow a wider tool wear span W. This in turn would increase the average loss per part associated with the resultant distribution. The loss function approach leaves the adjustment interval as before, since the additional loss due to wear is a function of W only, not variance. If the cost of adjustment

TABLE 1-2   Adjustment Interval Calculation

W, in	L_A	C_A	Total
.002	\$.013	\$.500	\$.513
.003	.030	.330	.360
.004	.053	.250	.303
.005	.083	.200	.283
.006	.120	.166	.286
.007	.163	.143	.306
.008	.213	.125	.338

were reduced, the goalpost approach would not require a change of the adjustment interval, whereas the loss function approach would change the adjustment interval to reduce the variation of the resultant distribution if the cost of adjustment were reduced.

This case study shows the economic penalty of allowing excess variation in a product or process. Taguchi uses a very comprehensive economic approach to on-line (production) quality control. A better description of the Taguchi methods would really be cost quality control.

1-8   Case Study: Automatic Transmissions*

Another example of the economic impact of excess variation occurred in the automatic transmission business with a major U.S. automobile company. Ford had contracted a Japanese supplier, Mazda, to make a certain portion of their front-wheel-drive automatic transmissions, with the balance of production made at a U.S. plant in Batavia, Ohio. Both sites were making transmissions to the same set of blueprints and the transmissions were being installed only in American cars. Mazda's version, as warranty records showed, had a substantially lower claim rate than the Batavia version. Ford investigated this phenomenon and found that Mazda's transmissions were made much more consistently than their own. On some critical control valve body components (valves, valve bores, and springs) which make a transmission shift automatically, Mazda was using only 27% of the allowed tolerance range, while Batavia was using 70%. Ford thought its plant was doing well, and by traditional standards it was, but the Mazda plant was superior. Not only were all the parts made to print, as were the U.S.-made parts, but they were more nearly like one another.

More investigation into the processes showed that Mazda was using a slightly more expensive and more complex grinder to finish the valve outer diameters. At first glance, one may think their parts were more expensive, but knowing that the loss function was at work, the parts were actually cheaper. The lower warranty bill for those transmissions substantiated that fact. By using this information, the Batavia plant has been able to improve its quality substantially and, in the first quarter of 1987, surpassed the Mazda level.

To summarize, continuous reduction of variation even within the allowed tolerance limits is a must to provide a more desirable product to the customer. This, in turn, is a more competitive product with a lower loss to society associated with it.

---

*Ford Motor Company, Dearborn, Michigan, 1987.

1-9   Factory Tolerances

One attribute of the loss function is to help determine what factory tolerances should be. For example, in automatic transmissions the shift points are supposed to occur at a certain speed at a certain throttle position. Heavy-duty truck users are particularly sensitive to this characteristic. Let's say that it costs the producer $100.00 to adjust a valve body under warranty when a customer complains of the shift point. From the information the engineers have available, the average customer would ask for an adjustment if the shift point is off from the nominal by 40 rpm transmission output speed on the first-to-second gear shift. The loss function is then

$$\text{Loss} = k(y - m)^2$$
$$\$100.00 = k(40)^2$$
$$k = \$0.0625/\text{rpm}^2$$

At the factory, the adjustment can be made at a much lower cost, approximately $10.00, which comprises the labor expense for time required to adjust the shift and rerun the test. What should be the limits that define when an adjustment should be made at the factory?

$$\text{Loss} = k(y - m)^2$$
$$\$10.00 = 0.0625(y - m)^2$$

$$(y - m)^2 = \left(\frac{\$10.00}{0.0625} \right) = 160$$

$$(y - m) = (160)^{-2} = \pm\ 13\ \text{rpm}$$

Therefore, if the transmission shift point is further than 13 rpm from the desired nominal, it is cheaper to adjust it at the factory than to wait for a complaint and subsequent adjustment under warranty. The cost is always lower to make it right at the factory than to find a poor-quality part when the customer has it in hand. Better yet, it always costs less to make it right at the point of production than to have to rework within the factory at a later time.

1-10   Other Loss Functions

The loss function can also be applied to product characteristics other than the situation in which the nominal value is the best value: where lower is better or higher is better, for instance. A good example of a lower-is-better characteristic is the waiting time for your order delivery at a fast-food restaurant. If the attendant tells you that it will be a moment for your hamburger to come up, then you sense some loss; the longer you have to wait, the larger the loss. Microfinish of a machined surface, friction loss, or wear are also examples of lower is better. Efficiency, ultimate strength, or fuel economy are examples of higher is better.

The loss function for a lower-is-better characteristic is shown in Fig. 1-10. The cost constant k can be calculated similarly to the nominal-is-best situation. There is some loss associated with a particular value of y. The loss can then be calculated for any value of y based on that value of k. This loss function is identical to the nominal-is-best

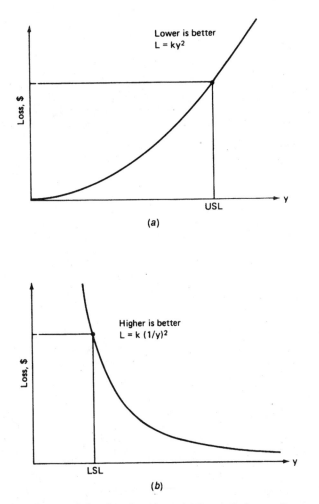

Figure 1-10   Other loss functions. (a) Lower is better, $L = ky^2$; (b) higher is better, $L = k(1/y^2)$.

TABLE 1-3   Types of Loss Functions

Type of characteristic	Loss for an individual part	Average loss per part in a distribution
HB	$k(1/y^2)$	$k[1/\bar{y}^2][1 + (3S^2/\bar{y}^2)]$
NB	$k(y - m)^2$	$k[S^2 + (\bar{y} - m)^2]$
LB	$k(y^2)$	$k[S^2 + (\bar{y}^2)]$

situation when $m = 0$, which is the best value for a lower-is-better characteristic (no negative values). This equation takes the form

$$L = k[S^2 + (\bar{y})^2]$$

The loss function for a higher-is-better characteristic is also shown in Fig. 1-10. Again, the cost constant can be calculated based upon some loss associated with a particular value of y. Subsequently, any value of y will have a loss assessed. The average loss per unit may be determined by finding the average value for $1/y^2$. This is mathematically equivalent to

$$L = k\left(\frac{1}{\bar{y}^2}\right)\left[1 + \left(\frac{3S^2}{\bar{y}^2}\right)\right]$$

Table 1-3 summarizes the different loss functions for the three types of characteristics, both for an individual part and for the average loss for a part from a distribution of parts.

1-11   General Loss Function for Nominal-Is-Best Situation

Returning once again to the general loss function for a nominal-is-best situation,

$$\text{Loss} = k[S^2 + (\bar{y} - m)^2]$$

one can see that the equation is made up of two parts: the variance and the relative location of the average of a performance characteristic of a group of products. Therefore, to minimize loss to society, the product characteristic needs to be centered at the nominal value and the variance of that characteristic needs to be reduced.

The loss function entails two aspects of quality management within a factory. First, the variance is the product and process engineer's job to establish before the start of production and to improve as time goes on (lower loss). Second, the centering of the distribution is the responsibility of the production (manufacturing) people on a day-in, day-out basis. This is off-line quality control (designers) and on-line quality control (makers), with reference to Fig. 1-1. The loss function is not intended to be accurate to the hundredth of a penny but is intended to show how reduced variation can result in reduced losses.

1-12   Summary

This chapter offered a different view of quality than has traditionally been used; uniformity of products is of greater concern than just conformance to specifications. Also considered was the economic impact of reducing variation in products and processes. Several examples were offered to show the relationship of variability to customer desires and a way of quantifying the value (loss function) of quality (reduced variation). The loss function addressed two descriptive statistics of a group of products: the average and the variance. The means to achieve a certain average from a product or process will not be addressed, since this is manufacturing's responsibility. Continuous reduction in the variance, however, can be intentionally achieved through the use of the methods covered in the remaining chapters of this text.

Problems

1-1   What is the k value for a pop-top tab for a beverage container? The cost to scrap is $0.02 if the tab is outside specifications and may fail to open the container. The specification is 1.1 to 1.3 pounds force.

1-2   Three processes make frictional clutch plates for an automatic transmission; none of the processes makes more than 0.27% scrap relative to the upper and lower specification limits. The drawing specifies an overall thickness of 0.125 in $\pm$ 0.010 in. The cost to scrap is $2.50 per part. What is the k value in the loss function?

1-3   What is the average loss per part for each of the groups of parts shown in Fig. 1-11? The population limits shown are assumed to be $\pm$ 3 sigma limits.

1-4   What conclusions are there concerning variation and the location of the average of the populations?

1-5   What is the k value for a situation concerning eccentricity of bearing journals on an engine crankshaft? The amount of eccentricity allowed is to be no more than 0.002 in (0.05 mm) on any one of the three center main journals with respect to the end two main journals. The value of a crankshaft at the final grinding operation is approximately $7.50.

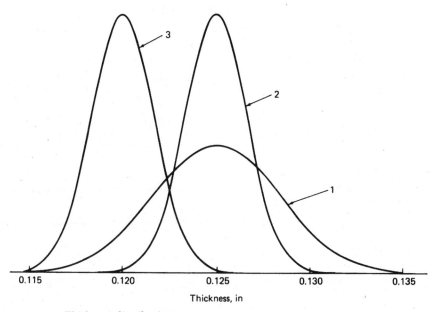

Figure 1-11   Thickness distributions.

1-6   What is the k value for a situation concerning the life of the rubber portion of windshield wipers? Life is measured in hours on an aggressive test consisting of thermal cycles, washing cycles, and wiping cycles. The end of useful life is determined when streaking appears during the washing/wiping cycles. The minimum acceptable on the test to provide satisfactory customer performance is 1000 hours. The value of a wiper blade insert is $1.00.

1-7   Referring to Prob. 1-5, what is the average loss per part if the average eccentricity in a production run is 0.0014 in (0.036 mm) and the standard deviation is 0.0002 in (0.005 mm)?

1-8   Referring to Prob. 1-6, what is the average loss per part if the average life is 1400 hours and the standard deviation is 30 hours?

1-9   Referring to Sec. 1-9, what would the new factory tolerances be if a better machine could set shift points for a cost of only $8.00?

2

The Design of Experiments Process

2-1 Introduction to the DOE Process

The primary purpose of this chapter is to provide a concise guide for executing an effectively designed experiment.

2-1-1 The purpose of experimentation

The purpose of product or process development is to improve the performance characteristics of the product or process relative to customer needs and expectations. The purpose of experimentation should be to understand how to reduce and control variation of a product or process; subsequently, decisions must be made concerning which parameters affect the performance of a product or process. The loss function quantifies the need to understand which design factors influence the average and variation of a performance characteristic of a product or process. By properly adjusting the average and reducing variation, the product or process losses can be minimized.

2-1-2 Basis of experimentation

This approach is based on the use of orthogonal arrays (Taguchi) to conduct small, highly fractional factorial experiments up to larger, full-factorial experiments. The use of orthogonal arrays is just one methodology to design an experiment, but probably the most flexible in accommodating a variety of situations and yet easy for nonstatistically oriented people to execute on a practical basis.

2-1-3 Definition of a designed experiment

A designed experiment is the simultaneous evaluation of two or more factors (parameters) for their ability to affect the resultant average or variability of particular product or process characteristics. To accomplish this in an effective and statistically proper fashion, the levels of the factors are varied in a strategic manner, the results of the particular test combinations are observed, and the complete set of results is analyzed to determine the influential factors and preferred levels, and whether increases or decreases of those levels will potentially lead to further improvement. It is important to note that this is an iterative process; the first round through the DOE process will many times lead to subsequent rounds of experimentation. The beginning round, often referred to as a screening experiment, is used to find the few important, influential factors out of the many possible factors involved with a product or process design. This experiment is typically a small experiment with many factors at two levels. Later rounds of experiments typically involve few factors at more than two levels to determine conditions of further improvement.

2-1-4 The design of experiments process

The DOE process is divided into three main phases which encompass all experimentation approaches. The three phases are (1) the planning phase, (2) the conducting phase, and (3) the analysis phase. The planning phase is by far the most important phase for the experiment to provide the expected information. An experimenter will learn something from any experiment; sometimes the information is in a positive sense and sometimes in a negative sense. Positive information is an indication of which factors and which levels lead to improved product or process performance. Negative information is an indication of which factors don't lead to improvement, but no indication of which factors do. If the experiment includes the real, yet unknown, influential factors and appropriate levels, the experiment will tend to yield positive information. If the experiment does not include the real influential factors, the experiment will yield negative information. The planning phase is when factors and levels are selected and, therefore, is the most important stage of experimentation. Also, the correct selection of factors and levels is nonstatistical in nature and is more dependent upon product and process expertise.

The second most important phase is the conducting phase, when test results are actually collected. If experiments are well planned and conducted, the analysis is actually much easier and more likely to yield positive information about factors and levels.

The analysis phase is when the positive or negative information concerning the selected factors and levels is generated based on the previous two phases. The analysis phase is least important in terms of whether the experiment will successfully yield positive results. This phase, however, is the most statistical in nature of the three phases of the DOE by a wide margin. Because of the heavier involvement of statistics, the analysis phase is typically the least understood by the product or process expert.

The major steps to complete an effective designed experiment are listed in the following text. The planning phase includes steps 1 through 9, the conducting phase is step 10, and the analysis phase includes steps 11 and 12.

1. State the problem(s) or area(s) of concern.

2. State the objective(s) of the experiment.

3. Select the quality characteristic(s) and measurement system(s).

4. Select the factors that may influence the selected quality characteristics.

5. Identify control and noise factors (Taguchi-specific).

6. Select levels for the factors.

7. Select the appropriate orthogonal array (OA) or OAs.

8. Select interactions that may influence the selected quality characteristics or go back to step 4 (iterative steps).

9. Assign factors to OA(s) and locate interactions.

10. Conduct tests described by trials in OA(s).

11. Analyze and interpret results of the experimental trials.

12. Conduct confirmation experiment.

These steps are fundamentally the same regardless of whether one is designing a Taguchi-based experiment or a classical design. The significant differences between the two approaches are in steps 7 through 9. All designed experiments require that a certain number of combinations of factors and levels be tested to observe the results of those test conditions. The Taguchi approach relies on the assignment of factors in specific orthogonal arrays to determine those test combinations. Taguchi is not the first to utilize an orthogonal array or equivalent approach to experimentation. All Latin square, Greco-Latin square, or Yates algorithm experiments are based on the same approach. Plackett and Burman utilized Hadamard matrices to perform similar experiments during World War II.

This process describes one loop through the DOE process. Two or more passes through the process are often utilized; earlier rounds of experimentation provide a growth of knowledge and a basis for later rounds of experimentation.

2-2   Task Aids and Responsibilities for DOE Process Steps

Table 2-1 delineates the task aids that facilitate moving through the DOE process steps and who is mainly responsible for successful completion of those tasks. The definitions for control and noise factors are provided in Sec. 2-3-5.

TABLE 2-1   DOE Process Tasks, Task Aids, and Responsibilities

Tasks	Task aids	Who
State problem(s)	Quality function deployment, test failures, warranty items, scrap items, Pareto analysis	Product and/or process experts
State objective(s)	Customer requirements, competitive benchmarks	
Select quality characteristic(s) & measurement system(s)	Gauge repeatability & reproducibility analysis	
Select factors and interactions; determine control and noise factors	Fishbone diagram, flowcharts, SPC charts	
Select levels	Specification limits, operational limits	
Select orthogonal array(s)	OA selection tables D-1, D-2, blank OAs	DOE expert
Assign factors & interactions to orthogonal array(s)	Assignment tables D-3, D-4; interaction tables; OA modification rules	
Conduct tests	Computer software, trial data sheets, randomization plan, part serialization plan, material logistics plan	Product, process, and DOE experts
Analyze and interpret data	Observation method, column effects method, ANOVA, computer software, plotting, ranking (magnitude & time, order)	DOE expert
Confirmation test	Estimates of the mean confirmation experiment flowchart	Product, process, and DOE experts

2-3 DOE Process Step Complete Description

The following sections of this chapter provide more information relevant to each of the DOE process steps. The initial steps in the planning phase, steps 1 through 6, are particularly emphasized. The final steps of the planning phase, steps 7 through 9, are covered in detail in Chap. 3. The conducting phase is covered in detail in Chap. 4, and the final phase of analysis and interpretation is covered in Chaps. 5 and 6.

The material in the following sections includes what information is needed to complete that step, what is accomplished by performing that step, what information will aid the successful completion of that step, pertinent comments relevant to that step, and two experimental examples of that particular step in the DOE process.

2-3-1 Step 1. State the problem(s) or area(s) of concern

What is needed to perform this step:

Data that characterizes the problem as it occurs

How the problem is observed

When the problem occurs

How severe the problem is

Where the problem occurs

(Why the problem occurs is not appropriate at this step)

What is accomplished by completing this step:

A statement is developed that clearly and concisely describes the problem (addresses the real problem, not symptoms of the problem).

Task aids:

Quality function deployment critical items

Product or process validation test failures

Scrap reports and warranty items

Products and causes that stop production

A problem can be viewed as any difference that exists between an ideal situation and the actual situation. For example, a process may currently make all products to specifications, which at first may not seem like a problem. However, the ideal situation would be to manufacture exactly identical products. Since there is variation present in the process, even though all products may be to specifications, there is a problem to be addressed, and that is to further reduce variation.

Increasing the technical accuracy of the problem statement enhances the effectiveness of later steps in the DOE process. A greater technical understanding of the problem improves the selection of the quality characteristic(s), the measuring system(s), the factor selection, and factor level selection. Increasing the technical accuracy of the problem is improved by addressing the real problem and not symptoms of the problem.

Examples include the following.

Water pump experiment.   A new engine design has water pump leaks on almost all of the initial assemblies at the splitline of the pump body at the gasket interface. The leaks are observed on the final engine assembly test stand during a standard production final test. Leakage severity ranges from a nonleaker to a severe leaker that may have several drips per minute of testing after stabilizing at operating temperature.

Die-cast piston experiment.   Die-cast pistons are low in hardness at times and make production machining difficult. The low-hardness pistons also have a lot of variation in hardness within a given piston. Low hardnesses range from 60 to 70 Rockwell B (R_B) hardness number; one piston may vary that much from the skirt to the dome, with the skirt having the higher hardness. Five to ten percent of the pistons are outside of the specification limits of 65 to 75 R_B.

2-3-2   Step 2. State the objective(s) of the experiment

What is needed to perform this step:

A problem statement

Competitive benchmark information concerning the problem

Customer information concerning the problem

What is accomplished by completing this step:

A determination of the required performance per customer requirements and competitive benchmarks

Task aids:

Customer requirements (internal or external)

Competitive benchmarks of similar products or processes

The statement of the experimental objective provides exit criteria for the experiment; how to determine when the experimental process should be stopped and other, more important problems are addressed. Examples include the following.

Water pump experiment. Determine the design or process parameters with appropriate nominals and tolerances that will eliminate water pump leaks entirely.

Die-cast piston experiment. Determine the alloy or process parameters with appropriate nominals and tolerances that will increase the average hardness to at least 75 points on the R_B scale (specifications will be changed if the results are higher than the current specification) and also reduce variation within a piston to less than 5 points on the R_B scale.

2-3-3 Step 3. Select the quality characteristic(s) and measurement system(s)

What is needed to perform this step:

A problem statement

An experimental objective

What is accomplished by completing this step:

A determination of the quality characteristic(s) to measure

A determination of the appropriate measurement system(s)

A determination of the people who will do the measurements

A determination of the method for measuring results

Task aids:

Process flow diagrams

Gage repeatability and reproducibility studies

The selection of quality characteristics to measure as experimental outputs greatly influences the number of tests that will have to be done to be statistically meaningful. Typically, quality characteristics that are variable (continuous) in nature require substantially fewer tests than quality characteristics that are attribute (discontinuous) in nature to achieve the same level of statistical significance. Semivariable (multiple classes of attribute) data is preferred over pure, two-class attribute data for the same reason.

Whenever it is possible to convert attribute data to a variable data, this should be done. For example, a beverage container that fails to open using the pop-top tab can be considered either an attribute or a variable data case. Attribute data considers that the container either opened successfully or the tab failed. Variable data considers the distribution of the strength of the tab versus the distribution of the strength of the top. Failures occur when the two distributions overlap slightly and the top is slightly stronger than the tab, as shown in Fig. 2-1. Since this type of failure is very infrequent, a large sample size will be required to determine the failure rate and what changes will reduce the failure rate in an attribute data viewpoint. In a variable data viewpoint, however, considerably fewer samples will be needed to determine what to change to either:

1. Slightly increase the average strength of the tab

2. Slightly decrease the average strength of the top

3. Decrease the variation of the tab or top strength

Any of these changes would reduce or eliminate the overlap of the distributions, thereby reducing or eliminating the container opening failures. Two separate experiments would be conducted: one for the factors that influence tab strength and another for factors that influence top strength.

Multiple quality characteristics may be evaluated in the same experiment. Each of the characteristics will require a separate analysis of results and perhaps an overall analysis.

It is recommended that one perform a measurement system repeata-

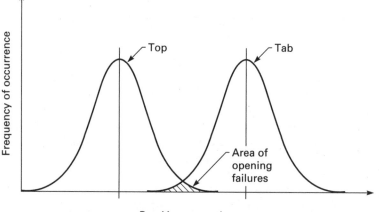

Figure 2-1 Beverage container tab and top strength comparison.

bility and reproducibility study prior to experimentation to understand the measurement system contribution to variation (applies to both types of data, variable or attribute).

Examples include the following.

Water pump experiment. Water pump leaks will be measured by observing the leakage and rating the leakage on a 0 (no leakage) to 5 (severe leakage) semivariable scale. Better measurement systems might be the number of drips in a 1-min time period or the weight of dripped fluid, which would be true variable data.

Die-cast piston experiment. Die-cast piston hardness will be measured on a Rockwell testing machine with a piston holding fixture adapted to the stand using the B scale.

2-3-4   Step 4. Select the factors that may influence the selected quality characteristic(s)

What is needed to perform this step:

A problem statement

Product design specifications

Process control plans

Process flow diagrams

Process routings

Statistical process control chart results

Product and process technical expertise (key information)

What is accomplished by completing this step:

A determination of a list of factors to be evaluated in the experiment for their effect on the selected quality characteristic(s)

Task aids:

Brainstorming with product and process technical experts

Process flow diagrams

Process routings

Product and process fishbone diagrams

Statistical process control charts

Product design specifications

Process control plans

This is the most important step of the DOE process. If important factors are unknowingly left out of the experiment, then the information gained from the experiment will not be in a positive sense. It will be information about which factors do not make a difference in the quality characteristic and other factors will have to be investigated. A recommended strategy is to focus on the real problem cause or failure mode and begin an investigation with many factors rather than just a few.

The determination of which factors to investigate hinges upon the product or process performance characteristic(s) or response(s) of interest. The customer who eventually uses a product expects or needs some function from a product. If during initial development stages of a product the function is not provided or consistently provided, the performance characteristic will have to be improved. Several methods are useful for determining which factors to include in initial experiments. These include brainstorming, flowcharting (especially for processes), and cause-effect diagrams.

Brainstorming. This activity involves bringing together a group of people associated with the particular problems and soliciting their advice concerning what to investigate. Here it is very appropriate to bring in product or process experts and statistically oriented people to discuss the factors and the structure of the experiment.

Flowcharting. In the case of a process, flowcharts are particularly useful in the determination of factors affecting the process results. The flowchart adds some structure to the thought process and thus may avoid the omission of important factors.

An example of a process flowchart for a casting problem is relatively simple. The casting process involves the steps shown in Fig. 2-2. The factors suggested from the flowcharts are

Pour casting:	Temperature of metal
	Speed of pouring
	Chemistry of metal
Cool in mold:	Time in mold
	Ambient temperature of mold
Shake-out casting:	Intensity of vibration
	Time of vibration
Air cool:	Ambient temperature
	Rate of air flow
Shot blast:	Intensity of shot blast
	Time of shot blast

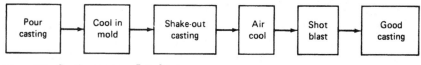

Figure 2-2 Casting process flowchart.

The problem observed at the end of the casting process is that a percentage of the castings are cracked at various locations. The factors to include in an experiment should be the ones thought relevant to the problem of casting cracks. All factors that are thought to influence a performance characteristic should be included in the initial round of experimentation. It is better to have many factors at few levels for early experiments. The purpose of the first stages of experimentation is to eliminate many factors from contention and find those important few factors that do contribute to a product problem or contribute to product quality improvement.

Cause-effect diagram. The structure for a cause-effect (C-E) diagram begins with the basic effect that is produced and progresses to what causes there may be for this effect. Primary, secondary, and perhaps tertiary causes are branched off the main trunk of the effect tree. The C-E diagram for the casting problem would appear as in Fig. 2-3.

Again, the selection of factors to be included in the experiment

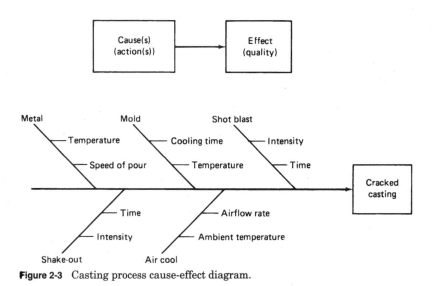

Figure 2-3 Casting process cause-effect diagram.

should depend on which might affect casting cracks. Ishikawa* provides several suggestions for developing C-E diagrams. After factors are selected, any interactions that are of interest should be noted.

Not a lot of emphasis will be placed on factor selection, because most engineers and scientists have specialized knowledge concerning their particular product and have, therefore, the best background for the selection process. What is needed is the proper structure of an experiment for the factors chosen. The design of experiments process is a universal tool that may be applied to a wide range of products and processes.

Examples include the following.

Water pump experiment. Factors which may influence the splitline performance would include such things as gasket design, cover design, bolt torques, torque sequence of bolts, gasket sealant, and pump cover finish. Factors which influence part porosity leaks or shaft seal leaks are irrelevant according to the original problem statement.

Die-cast piston experiment. Factors which may influence piston hardness would be processing factors such as metal temperature, die water cooling, cycle time, and air cooling after ejection. Alloying factors may not have a strong effect since there is considerable variation within a piston and a given piston is made up of a rather homogeneous volume of aluminum. The skirt being harder than the dome provides a clue relating to possible thermal or cooling effects and, therefore, the need to include temperature-related factors.

2-3-5　Step 5. Identify control and noise factors

What is needed to perform this step:

A list of factors to evaluate for their effect on the quality characteristic of interest

What is accomplished by completing this step:

The list is separated into control and noise factors

Task aids:

Control and noise factor definitions

Control factors are those factors that a manufacturer can control in the design of a product, the design of a process, or during a process.

---

*K. Ishikawa, *Guide to Quality Control*, Chap. 3, Asian Productivity Organization, Tokyo, 1976.

Noise factors are those things that a manufacturer cannot or wishes not to control for cost reasons. Noise factors may be controlled temporarily during an experiment, but in an actual production or customer environment may not or cannot be controlled at all. Because noise factors are either too expensive or impossible to control on a continuous basis, the focus of the experiment should be on the effects of the true control factors. Also, Chap. 7, concerning the concept of parameter design, should be referenced for another approach to experimenting with control and noise factors.

Examples of separating factors into groups of control and noise factors for the water pump and die-cast piston experiments are shown in Table 2-2.

2-3-6   Step 6. Select levels for the factors

What is needed to perform this step:

A list of control and noise factors

Product or process technical expertise

Product or process specification or operating limits

What is accomplished by completing this step:

A determination of the values for all the levels of the selected factors and the number of levels for each factor

Task aids:

Product or process technical expertise (key information)

Product or process specification or operating limits

A minimum of two levels are required to evaluate a factor's effect on a given quality characteristic. When screening (beginning round)

TABLE 2-2   Control Factor and Noise Factor Examples

	Control factors	Noise factors
Water pump experiment	Cover design, gasket design, bolt torques, assembly method	Assembly persons, time after assembly
Die-cast piston experiment	Metal temperature, die cooling amount, air cooling amount, cycle time	Position on a piston, metal batches, different sets of dies, ambient temperature
Other examples	Materials, speeds, feeds, diameters, tire diameter, tread design	Batches of materials, machine operators, ambient temperature, ambient humidity, road surface, vehicle speeds

experiments are done and several factors are under consideration, then it is recommended that the experiments use only two levels where possible to keep the size of experiment to a minimum. The intent of a screening experiment is to identify the few factors out of the many possible factors that actually have a substantial effect on the quality characteristic of interest. In later experiments, when important factors have been isolated, then more levels may be evaluated without the penalty of using a large experiment. When multiple levels are utilized, the levels should be equally spaced (equal intervals between levels) for analytical reasons, as discussed in Sec. 5-7. These recommendations apply for levels of continuous factors such as temperature, pressure, speed, and time. However, when discrete factors, such as different materials, are considered, then all levels (three different materials, for example) may have to be tested in the screening experiment. Discrete parameters may assume only particular values such as off or on; material A, B, or C; or engine cylinder number 1, 2, 3, or 4.

If continuous parameters are being used, then the initial experiment should be at two levels only; interpolation or extrapolation may be used to predict the results at other levels. If discrete factors are used, then interpolation or extrapolation may be meaningless. For instance, when the use of three different materials is possible, there is no way to interpolate or extrapolate in order to predict results of a fourth possible material. If discrete parameters are studied, then more than two levels may be required in initial experiments.

Product or process technical expertise is the single most important source for the selection of appropriate values for the factor levels. Any factor can be made to look insignificant by choosing levels that are too close together and, conversely, any factor can be made to look significant by choosing levels that are too far apart. The levels need to be in an operational range of the product or process. The issue at hand is which of the selected levels, if any, cause some improvement in the quality characteristic.

The numbering scheme for the levels (which level is identified as first or second) is not critical, and two approaches may be considered. For continuous factors, the first level may be the lower of the two values being tested and the second level the higher level. This provides an intuitive relationship between the value of the factor and the level designation. Another approach, especially in a process development situation, is to assign the first level to all the factor values that represent the current operating conditions. With this approach, one of the trials in the experiment will automatically represent the baseline conditions.

Examples of selecting levels for the chosen factors in the water pump and die-cast piston experiments are shown in Table 2-3.

TABLE 2-3   Factor Level Examples

	Factors	Level 1	Level 2
Water pump leak experiment	A cover design	Production	New
	B gasket design	Production	New
	C front bolt torque	LSL	USL
	D sealant	No	Yes
	E pump finish	Rough	Smooth
	F back bolt torque	LSL	USL
	G torque sequence	Front-back	Back-front
Die-cast piston experiment	A copper %	LSL	USL
	B magnesium %	LSL	USL
	C zinc %	LSL	USL
	D die cooling	On	Off
	E air cooling	On	Off
	Z piston position	Dome	Skirt

2-3-7   Step 7. Select the appropriate orthogonal array(s)

What is needed to perform this step:

The number of control and noise factors

The number of levels for the specific factors

The experimental resolution desired

What is accomplished by completing this step:

The determination of the appropriate orthogonal array(s) for the experiment

Task aids:

Orthogonal array selection tables

See Chap. 3, "Orthogonal Array Selection and Utilization," for complete details of this step in the DOE process.

2-3-8   Step 8. Select interactions that may influence the selected quality characteristics or go back to step 4 (iterative steps)

What is needed to perform this step:

The OA column numbers which have no factors assigned to them

What is accomplished by completing this step:

A list of interactions or additional factors to evaluate

Task aids:

Product or process technical expertise

See Chap. 3 for complete details of this step in the DOE process.

2-3-9   Step 9. Assign factors to OA(s) and locate interactions

What is needed to perform this step:

A list of selected control and noise factors

A list of interactions that may be of interest

The number of levels for the factors

The existence or nonexistence of a nesting factor

The existence or nonexistence of combined factors

The existence or nonexistence of idle column factors

What is accomplished by completing this step:

An effective assignment of the selected factors to an OA or OAs (modified or unmodified)

Task aids:

OA assignment tables

Blank OAs (standard and modified)

OA interaction tables

OA modification techniques and rules

See Chap. 3 for complete details of this step in the DOE process.

2-3-10   Step 10. Conduct tests described by trials in OA(s)

What is needed to perform this step:

An assignment of the selected factors and levels to an OA or OAs (modified or unmodified)

Values for all the levels of the factors

Trial data sheets

Determination of the sample size to be used

A randomization strategy

A material logistics strategy

A data collection strategy

People roles and responsibilities identified for the duration of the experiment

Test equipment availability

What is accomplished by completing this step:

A complete set of test data is generated for each of the trials in the OA(s)

Task aids:

Trial data sheets

Computer software to generate trial data sheets

Sample size tables

Randomization strategy table

See Chap. 4, "Conducting Tests," for complete details of this step in the DOE process.

2-3-11   Step 11. Analyze results of the experimental trials

What is needed to perform this step:

A complete set of test results for each and every trial of the experiment and the test order of the trials or repetitions

What is accomplished by completing this step:

A determination of factors which are thought to be influential in changing the average result or variation of the results

A determination of how influential other uncontrolled factors may have been during the experiment

A determination of which levels of the influential factors are the most desirable from a technical viewpoint

A determination of the expected results of the optimum combination of influential factors and levels

A determination of whether further improvement in results is possible

Task aids:

Observation method

Column effects method

Analysis of variance and percent contribution method

Statistical F ratio tables

Plotting of influential factors

Ranking of test results

Plotting in test order

Computer software to do analysis and plots

Data transformation techniques

See Chap. 5, "Analysis and Interpretation Methods for Experiments," for complete details of this step in the DOE process.

2-3-12   Step 12. Conduct confirmation experiment

What is needed to perform this step:

Identification of statistically significant factors and the levels that are most desirable from a technical viewpoint

Identification of statistically insignificant factors and the levels that are most economical

What is accomplished by completing this step:

A validation or invalidation of the interpretation of the significant and insignificant factors and levels to achieve the expected results

Task aids:

Mean estimate

Confidence interval

Confirmation experiment trial data sheet

Confirmation experiment flowchart

See Chap. 6, "Confirmation Experiment," for complete details of this step in the DOE process.

2-4   Summary

This chapter introduced the fundamental steps in performing effective designed experiments utilizing Taguchi orthogonal arrays to simultaneously evaluate several factors for their effect on product or process performance. The following chapters delve into more detail on the DOE process steps.

Problems

2-1   What are the three main phases of experimentation?

2-2   What are the five most important steps in the planning phase?

2-3   What is the difference between positive and negative information produced by an experiment?

2-4   What are factors in a designed experiment?

2-5   What is the difference between control and noise factors?

2-6   What are factor levels in a designed experiment?

2-7   What purpose do flowcharts and cause-effect diagrams serve?

2-8   Give some examples of variable characteristics.

2-9   Give some examples of attribute characteristics.

3

Orthogonal Array Selection and Utilization

This chapter deals with the final steps of the critical planning phase of the DOE process. These three steps accomplish one main objective which is to determine the combinations of factors and levels to test to estimate the effect of those factors. This can be accomplished in an effective manner with the proper testing strategy. The recommended strategy is discussed specifically in Sec. 3-3 and summarized in Sec. 3-4.

3-1 Typical Test Strategies

Engineers and scientists are most often faced with two product (or process) development situations. The terms product and process can be used interchangeably in the following discussion, because the same approaches would apply if developing either a product or process. One development situation is to find a parameter that will improve some performance characteristic to an acceptable or optimum value. A second situation is to find a less expensive, alternative design, material, or method which will provide equivalent performance. Depending on which situation the experimenter is facing, different strategies may be used. The first problem of needing to improve performance is the most typical situation.

When searching for improved or equivalent designs, the person typically runs some test, observes some performance of the product, and makes a decision to use the new design or to reject the new design. It is the quality of this decision that can be improved upon when proper test strategies are utilized, in other words, to avoid either the mistake of using an inferior design or not using an improved design.

Before the discussion of orthogonal arrays (OAs), it would be best to review some often-used test strategies. Not being aware of efficient, proper test strategies, experimenters resort to the following approaches. The most common test plan is to evaluate the effect of one parameter on product performance. A typical progression of this approach, when the first parameter chosen doesn't work, is to evaluate the effect of several parameters on product performance one at a time. The most urgent and desperate of situations finds the experimenter usually evaluating the effect of several parameters on performance all at the same time.

These different test strategies can be symbolized in the following manner. The simplest case of testing the effect of one parameter on performance would be to run a test at two different conditions of that parameter, for example, the effect of cutting speed on the microfinish of a machined part. Two different cutting speeds could be used and the resultant microfinish measured to determine which cutting speed gave the most satisfactory results. If the first level, the first cutting speed, is symbolized by a 1 and the second level, the second cutting speed, is symbolized by a 2, the experimental conditions would appear as in Table 3-1.

The * symbolizes the values for the microfinish that would be obtained on the different test samples. The sample of 2, for example only, under trial 1 could be averaged and compared to the average of the sample of 2 under trial 2 to estimate the effect of cutting speed. To do this in a statistically proper fashion, a valid number of samples under trial 1 and trial 2 would have to be made; two under each condition may not be adequate. Most engineers are not familiar with the statistical methods of determining the proper sample sizes, but there are statistics books that can aid in that determination.*†

If the first factor chosen fails to produce the hoped for results, the person usually resorts to testing some other factor, and the resultant test program would appear as shown in Table 3-2. In generic terms, let's assume the experimenter has looked at four different factors labeled *A, B, C,* and *D,* each evaluated one at a time. One can see in

TABLE 3-1   One-Factor Experiment

Trial no.	Factor level	Test results
1	1	* *
2	2	* *

---

*W. J. Diamond, *Practical Experiment Designs for Engineers and Scientists,* Lifetime Learning Publications, Belmont, Calif., 1981.

†C. Lipson and N. J. Sheth, *Statistical Design and Analysis of Engineering Experiments,* McGraw-Hill, New York, 1973.

TABLE 3-2   Several Factors, One at a Time

Trial no.	Factor and factor level				Test results
	A	B	C	D	
1	1	1	1	1	* *
2	2	1	1	1	* *
3	1	2	1	1	* *
4	1	1	2	1	* *
5	1	1	1	2	* *

Table 3-2 that the first trial is the baseline condition. The results of trial 2 can be compared to trial 1 to estimate the effect of factor A on product performance. The results of trial 3 may be compared to trial 1 to estimate the effect of factor B on product performance, and so on. Each factor level is changed one at a time, holding all others constant. This is the traditional "scientific" approach to experimentation often taught in today's high school and college chemistry and physics classes.

The third and most urgent situation finds the person in a desperate state and changing several things all at the same time in hopes that at least one of the changes will improve the results sufficiently. Again, one can see in Table 3-3 that the first trial represents the baseline condition. The average of the data under trial one may be compared to the average of the data under trial 2 to determine the combined effect of all factors.

All the methods have some type of limitation(s). These will be discussed separately.

3-1-1   One-factor experiment

The one-factor experiment evaluates the effect of one parameter on performance while ostensibly holding everything else constant. If there happens to be an interaction of the factor studied with some other factor, then this interaction cannot possibly be observed. Also, a one factor experiment doesn't use the test data in an effective manner. The statistically valid sample size may be eight tests for each level to detect a difference in performance between the levels that is technically important. With a total of 16 tests and a properly structured experi-

TABLE 3-3   Several Factors, All at the Same Time

Trial no.	Factor and factor level				Test results
	A	B	C	D	
1	1	1	1	1	* *
2	2	2	2	2	* *

ment that will be described later in this chapter, as many as three factors could have been evaluated with the same statistical sensitivity.

3-1-2   Several factors, one at a time

The main limitation of several factors, one at a time, is that no interaction among the factors studied can be observed. Also, this strategy makes limited use of the test data when evaluating factor effects. Of the 10 data points represented, only two are used to compare against two others; the remaining six data points are temporarily ignored. If an attempt is made to use all the data points, then the experiment will not be orthogonal. Orthogonality means that factors can be evaluated independently of one another; the effect of one factor does not bother the estimation of the effect of another factor. One provision of orthogonality is a balanced experiment: an equal number of samples under the various treatment conditions (an equal number of tests under A_1 and A_2).

For instance, the nonorthogonality of the data set in Table 3-2 can be seen. If all the data under level A_1 is averaged and all the data under level A_2 is averaged and compared, this is not a fair comparison of A_1 to A_2. Of the four trials under level A_1, three were at level B_1 and one at level B_2. The one trial under level A_2 was at level B_1. Therefore, one can see that if factor B has an effect on performance it will be part of the observed effect of factor A and vice versa. Only when trial 1 is compared to other trials, one at time, are the factor effects orthogonal.

3-1-3   Several factors, all at the same time

This situation makes separation of any of the main factor effects impossible, let alone any interaction effects. Some factors may be making a positive contribution and others a negative contribution, but no hint of this fact will exist. How can poor utilization of test data and a nonorthogonal situation be avoided? How can the interactions be estimated and still have an orthogonal experiment? The use of full-factorial experiments is one possibility. The use of some strategic orthogonal arrays is another.

3-2   Better Test Strategies

A full-factorial experiment can be symbolized in Table 3-4. One can see that the full-factorial experiment is orthogonal in this case. There is an equal number of test data points under each level of each factor. Note that under level A_1 factor B has two data points under B_1 condition and two under B_2 condition. The same is true under the level A_2. The same balanced situation is true when looking at the experiment

TABLE 3-4 Full-Factorial Experiment

| Trial no. | Factor and factor level | | y data (R_B) |
	A	B	
1	1	1	76 78
2	1	2	77 78
3	2	1	73 74
4	2	2	79 80

with respect to the two conditions of B_1 and B_2. Because of this balanced arrangement, factor A does not influence the estimate of the effect of factor B and vice versa. In matrix algebra, there are some mathematical relationships that are true for an experiment that is orthogonal, but the most practical view of this is to observe the balanced treatments within the experiment.

One can see that all possible combinations of the two factors and the two levels are represented in the preceding test matrix. Using this information, both the factor and interaction effects can be estimated. A full-factorial experiment is acceptable when only a few factors are to be investigated, but not very acceptable when there are many factors. If a full-factorial experiment is used, there is a minimum of 2^f possible combinations that must be tested (f = the number of factors each at two levels). Frequently, a typical engineering investigation may initially involve five or more factors.

The water pump leak problem described in Sec. 2-3-6 and the list of seven factors at two levels in Table 2-3 was an actual experiment performed at an engine plant. If a full-factorial experiment were to be used in this situation, then a total of 128 tests must be conducted. This type of experiment, shown in Fig. 3-1, estimates all the main fac-

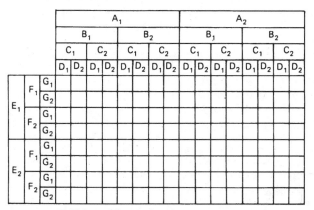

Figure 3-1 Full-factorial experiment.

tor effects and all the possible interactions, all orthogonal to one another. However, usual time and financial limitations preclude the use of a full-factorial experiment. How can an engineer efficiently (economically) investigate these design factors?

3-3   Efficient Test Strategies

Statisticians have developed more efficient test plans, which are referred to as fractional-factorial experiments (FFEs). FFEs use only a portion of the total possible combinations to estimate the main factor effects and some, not all, of the interactions. Shown in Fig. 3-2 are a $\frac{1}{2}$ FFE, a $\frac{1}{4}$ FFE, and a $\frac{1}{8}$ FFE. Certain treatment conditions are chosen to maintain the orthogonality among the various factors and interactions. It is obvious that a $\frac{1}{16}$ FFE with only eight test combinations, as shown in Fig. 3-3, is much more appealing to the experimenter from a time and cost standpoint. The information generated in such a small experiment is, however, substantially reduced from that of a full-factorial experiment.

Taguchi has developed a family of FFE matrices which can be utilized in various situations. In this situation, one possible matrix is an eight-trial OA, which is labeled an L8 matrix. An L8, two-level matrix is shown in Table 3-5.

Actually, this is a $\frac{1}{16}$ FFE, equivalent to Fig. 3-3, which has only 8 of the possible 128 combinations represented. One can observe seven columns in this array, similar to an earlier array (several factors one at a time), which may have a factor assigned to each column. When all columns are assigned a factor, this is known as a saturated design. The levels for the particular trials are designated by 1s and 2s as before. When water pump factors A through G are assigned to columns 1 through 7, there are eight unique pump assemblies described by all the different trials. The eight trial combinations match the eight descriptions in the $\frac{1}{16}$ FFE in Fig. 3-3. These are two different ways to describe exactly the same experiment; however, with the use of OA approach it is much easier to determine the appropriate orthogonal combinations and to perform the analysis as described in Chap. 5.

It is easy to see that all columns in an L8 OA provide four tests under the first level of the factor and four tests under the second level of the factor. This is one of the features that provides the orthogonality among all the columns (factors). The real power in using an OA is the ability to evaluate several factors in a minimum of tests. This is considered an efficient experiment since much information about factors is obtained from a few trials.

OAs were a mathematical invention recorded as early as 1897 by Jacques Hadamard, a French mathematician. The utility of these

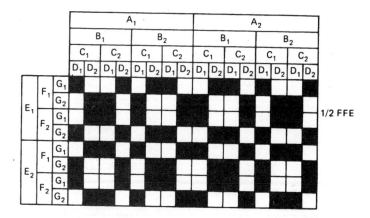

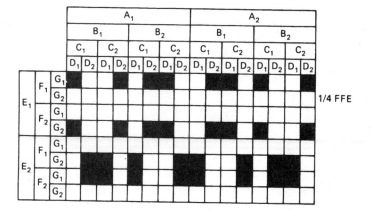

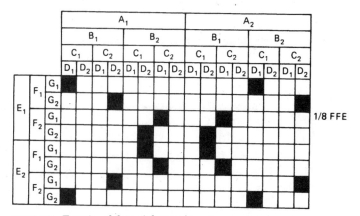

Figure 3-2 Fractional-factorial experiments.

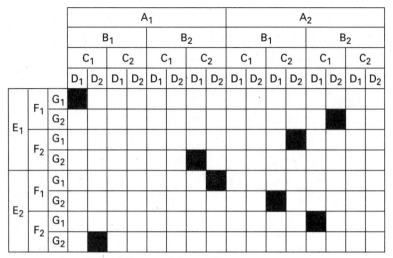

Figure 3-3 One-sixteenth fractional-factorial experiment.

TABLE 3-5   L8 OA Matrix

Trial no.	Column no.						
	1	2	3	4	5	6	7
1	1	1	1	1	1	1	1
2	1	1	1	2	2	2	2
3	1	2	2	1	1	2	2
4	1	2	2	2	2	1	1
5	2	1	2	1	2	1	2
6	2	1	2	2	1	2	1
7	2	2	1	1	2	2	1
8	2	2	1	2	1	1	2

arrays was not explored until World War II by Plackett and Burman, British statisticians, who employed the saturated approach previously described. The Hadamard matrices are identical mathematically to the Taguchi matrices; the columns and rows are rearranged. The assignment of factors to a saturated FFE is not difficult; all columns are assigned a factor. However, experiments which are not fully saturated may be more complicated to design.

3-3-1   Selection of the orthogonal array(s)

The selection of which OA to use predominantly depends on these items in order of priority:

1. The number of factors and interactions of interest
2. The number of levels for the factors of interest
3. The desired experimental resolution or cost limitations

The first two items determine the smallest orthogonal array that it is possible to use, but this will automatically be the lowest-resolution, lowest-cost experiment. The experimenter may choose to run a larger experiment (larger orthogonal array) which will have higher resolution potential but will also be more expensive to complete.

Orthogonal arrays. Two basic kinds of OAs are listed in the appendixes. Appendix B contains two-level arrays:

L4   L8   L12   L16   L32

Appendix C contains three-level arrays:

L9   L18   L27

The number in the array designation indicates the number of trials (different possible test combinations) in the array; an L8 has eight trials and an L27 has 27 trials, for example. A factor may be assigned to any and all columns in the OA. The 1s, 2s, and 3s within the trials of the OAs designate the appropriate level of the factor assigned to that column to be used for that specific trial.

Selection of OA. The number of levels used in the factors should be used to select either two-level or three-level types of OAs. If the factors are two-level, then an array from App. B should be chosen; if factors are three-level, then an array from App. C should be chosen. If some factors are two-level and some three-level, then the greater quantity should indicate which kind of OA is selected. Sections 3-5, 3-6, and 3-7 discuss how to modify basic OAs to accommodate a mixture of two-, three-, and four-level factors.

Table D-1 in App. D lists many of the options that are available in a two-level situation and Table D-2 lists those for a three-level situation. Referring to Table D-1, the numbers of two-level factors are listed across the top of the table and the possible OAs down the left-hand side. Looking below the number of two-level factors that may be under consideration for a given experiment, the possible OAs are indicated along with the maximum resolution possible. In screening (beginning-round) experiments, the recommended strategy is to start with the smallest OA that will accommodate the typically large number of factors under evaluation. This means that the resolution will be low in the first round of experiments (a small fractional factorial) and will progress to higher-resolution experiments (a large fractional or full factorial) as a few factors are identified as influential. This strategy will minimize the total number of tests to be conducted yet will yield meaningful information at the same time. If tests may be very inexpensively conducted, then a high-resolution, full-factorial experiment may be in order.

Once the appropriate OA has been selected, the factors can be assigned to various columns of the array and subsequent interaction columns located.

3-3-2   Interaction effect of factors

Pairs of factors in an experiment may interact with one another to provide a synergistic effect on a quality characteristic being studied. A typical interaction plot appears as in Fig. 3-4. The estimate of factor B's effect would depend upon which level of factor A was being used. At the A_1 level, the B effect appears to be relatively small and at the A_2 level, the B effect appears to be relatively large. It can also be stated that the estimate of factor A's effect depends upon which level of factor B was being used. At the B_1 level, the A effect appears to be relatively small and at the B_2 level, the A effect appears to be relatively

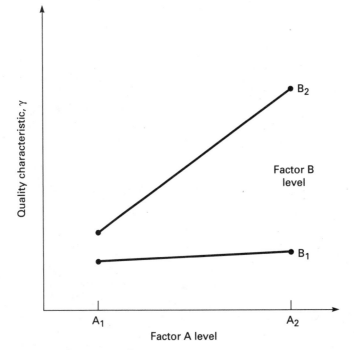

Figure 3-4 Interaction plot.

large. This mutual dependency between the factors is an interaction of those factors. When a two-factor experiment is conducted, there are three items that may be statistically estimated:

1. Factor A's overall effect to change the result
2. Factor B's overall effect to change the result
3. The interaction effect of factors A and B to change the result

Statistically, these are treated as three separate items which may have their individual strengths estimated.

There may be factors in an experiment that are thought to have interaction potential which may be of specific interest to the experimenter. This is an iterative step; the smallest OA that will accommodate the number of factors may have columns with no factors assigned. This happens because OA size (number of trials and columns) does not increase in increments of one, but in geometric progression with respect to the number of levels in the OA, for example, L4, L8, L16, L32, etc. Therefore, a list of 12 factors to be evaluated will require an L15 and will, subsequently, have three columns with no factors assigned which could be used to evaluate specific interactions or three additional factors with no increase in the size of the experiment. As a general recommendation, it is preferred to study more factors than to study interactions.

3-3-3 Assignment of factors and location of interactions

Before getting into the detail of using some methods of assigning factors and interactions, a discussion of a mathematical property of OAs is in order. OAs have several columns available for the assignment of factors and some columns will, subsequently, estimate the effect of interactions of those factors. The mathematical property of the OA works like this. If one factor is assigned to any particular column in a two-level array and a second factor is assigned to any other particular column, a specific third column will automatically have the interaction of those factors assigned to that column. The strength of factor A will be evaluated using one column, the strength of factor B will be evaluated using the second column, and the strength of the interaction of factor A and B will be evaluated using a specific third column. The pattern of which columns will be interaction columns is known for all of the orthogonal arrays.

Taguchi has provided two tools to aid in the assignment of factors to arrays and location of interactions in arrays:

1. Interaction tables

2. Linear graphs

Each OA has an interaction table and a particular set of linear graphs associated with it. The interaction tables contain all the possible interactions between factors (columns). The linear graphs indicate various columns to which factors may be assigned and which columns subsequently evaluate the interaction of those factors. The linear graphs are simply a visual representation of a portion of the interaction table. The interaction table is the master description of all the interaction columns in that array. Also, listed in Apps. B and C are the interaction tables for the two-level and three-level arrays, respectively.

Interactions for two-level OAs. The simplest OA, an L4, has a interaction table as shown in Table 3-6. Recall that the L4 OA has four trials and three columns. The first factor assigned to an OA may actually be placed in any column—column 2, as an example. The second factor may be assigned to any other column—column 3, as an example. If factor A is assigned to column 2 and factor B is assigned to column 3, the interaction table indicates that the $A \times B$ interaction will be in column 1, as shown in Table 3-7. The interaction table shows that the three columns are mutually interactive; 1 and 2 interact in 3, 2 and 3 interact in 1, and 1 and 3 interact in 2. Any assignment of factors A and B is mathematically and statistically equivalent. The interaction table for an L4 is actually a small portion of a larger interaction table that will handle up to an L32 OA (except for an L12 which does not have specific interaction columns). The interaction table for up to an L32 OA is shown in App. B.

TABLE 3-6   Two-Level Interaction Table

Column	2	3
1	3	2
2		1

TABLE 3-7   Two-Level Interaction Table with Factors

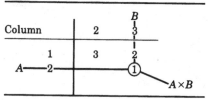

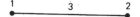

Figure 3-5 L4 linear graph.

An L4 OA has a linear graph that appears in Fig. 3-5. The linear graph indicates that factor A may be assigned to column 1, factor B to column 2, and the $A \times B$ interaction subsequently located in column 3. The dot represents a column available for a two-level factor and the line represents a column which will evaluate the interaction of the factors assigned to the respective dots. All of the two-level OA interaction tables and triangular tables function in this same manner. Taguchi emphasizes the use of linear graphs over the use of interaction tables; however, the linear graph assignment may cause resolution to be lower than optimum for a given-size OA.

Interactions for three-level OAs. The interaction table for an L9 appears in Table 3-8. The table again indicates all the possible interacting column relationships (pairs of columns) that evaluate the interaction. All the three-level OAs and triangular tables are used in the same manner. An interaction table for up to an L27 OA is shown in App. C.

Resolution of experiments. As was discussed earlier, Plackett and Burman's use of OAs resulted in the assignment of factors to all columns. Obviously, many interactions are confounded (mixed) with the main effects. This is the major compromise of using FFEs—to reduce the number of tests, some information must be surrendered.

An L8 OA provides a complex enough array to demonstrate the amount of confounding that may occur in an experiment. It is obvious that, if two factors are assigned to an L8, columns 1 and 2 are typically used and has the $A \times B$ interaction automatically located in column 3. What happens as more factors are added to the experiment? If three factors are assigned (A, B, and C), the assignment of factor C should be to column 4, which is the next truly empty column. According to the interaction table for two-level OAs, column 5 will have the $A \times C$ interaction, column 6 will have the $B \times C$ interaction, and column 7 will have the $A \times B \times C$ interaction. The interaction

TABLE 3-8 Three-Level Interaction Table

Column	2	3	4
1	3,4	2,4	2,3
2		1,4	1,3
3			1,2

table locates the $A \times B \times C$ three-factor interaction by finding the interaction of factor A and the $B \times C$ interaction in columns 1 and 6, respectively. Other combinations indicate the same factor and interaction columns:

Interactions $A \times (B \times C) = A \times B \times C$
Column numbers   1   6   7
Interactions $B \times (A \times C) = A \times B \times C$
Column numbers   2   5   7
Interactions $C \times (A \times B) = A \times B \times C$
Column numbers   4   3   7

The resultant column assignments for the factors and the interactions of this experiment are shown in Table 3-9.

In this situation, all main effects and all interactions can be estimated, which results in a high-resolution experiment. Resolution power indicates the clarity with which individual effects of factors and interactions may be seen (evaluated) in an experiment. Table D-1 in App. D indicates this experiment to be of resolution 4, which is also a full-factorial experiment in this case.

If another factor D is added to this experiment, the most logical column assignment is 7, automatically confounding D with a three-factor interaction. Any other choice confounds D with a two-factor interaction. The three-factor interaction is much less likely to occur, and if it does, then it will most likely be of smaller magnitude than a main effect or a two-factor interaction. Now the factors and interactions appear as in Table 3-10. The resolution power of the experiment is subsequently much lower, resolution 2, because of the greater amount

TABLE 3-9   Resolution 4 Experiment

			Column no.			
1	2	3	4	5	6	7
A	B	$A \times B$	C	$A \times C$	$B \times C$	$A \times B \times C$

TABLE 3-10   Resolution 2 Experiment

			Column no.			
1	2	3	4	5	6	7
A	B	$A \times B$	C	$A \times C$	$B \times C$	$A \times B \times C$
$B \times C \times D$	$A \times C \times D$	$C \times D$	$A \times B \times D$	$B \times D$	$A \times D$	D

of confounding in the columns. Also, the four-factor interaction, $A \times B \times C \times D$, cannot be estimated in this experiment because factor D was intentionally confounded with the $A \times B \times C$ interaction.

If a fifth factor is added, columns 3, 5, and 6 are available; then the resolution power will drop off further to a resolution 1 because of a factor being confounded with a two-factor interaction. Since factor E was confounded with interactions $A \times B$ and $C \times D$, the three-factor interactions $A \times B \times E$ and $C \times D \times E$ cannot be evaluated. This situation is shown in Table 3-11. This confounding of groups of two-factor interactions can be easily seen in an L8 OA, but occurs at a much higher degree of complexity as more factors are assigned to larger OAs. Also, in the previous examples, as one added factor is confounded with a three-factor interaction, all other factors are automatically mixed with a different three-factor interaction. As one added factor is mixed with a two-factor interaction, all other factors are automatically confounded with different two-factor interactions.

Table D-3 in App. D lists the column assignments that should be used to provide the highest resolution possible for two-level factors in standard OAs. If four or fewer factors are evaluated, then columns 1, 2, 4, and 7 of an L8 should have factors assigned to them. If eight factors are evaluated, then the columns indicated under an L16 should have factors assigned to them. The key thing to remember is that, as factors are added to a given OA, they should be placed in columns in which the lowest-order interaction is still of higher order than other columns. In the previous example (refer to Table 3-10), factor D was added into the three-factor interaction column of the experiment. Using this approach protects the resolution power of the experiment as factors are added.

Taguchi does not place much emphasis on the confounding that exists in a low-resolution experiment; however, it is good for the experimenter to be aware of such situations. A list of factors and interactions of interest might include factors A, B, and C plus interactions $A \times B$ and $A \times C$. These can be allocated to separate columns and they appear to be completely independent from any confounding, but this is not the case. Taguchi views interactions as being of mini-

TABLE 3-11   Resolution 1 Experiment

			Column no.			
1	2	3	4	5	6	7
A	B	$A \times B$	C	$A \times C$	$B \times C$	$A \times B \times C$
$B \times C \times D$	$A \times C \times D$	$C \times D$	$A \times B \times D$	$B \times D$	$A \times D$	D
$B \times E$	$A \times E$	E	$D \times E$	$A \times D \times E$	$B \times D \times E$	$C \times E$
				$B \times C \times E$	$A \times C \times E$	

mal interest because to utilize the interactive effect the experimenter must control two main effects. Since one or more main effects usually need to be controlled for a product or process anyway, the interaction causes no additional complications. In the case of an HB or LB characteristic, the main effects will indicate the proper combination of levels when the interaction effect is less than either main effect; therefore, knowledge of the interaction is irrelevant. The experimenter needs to know about the interaction only to be able to obtain the desired average response.

From a very practical point of view, all factors assigned to an experiment will not be equally influential in changing the average response. Therefore, what few factors are significant will have been evaluated as if the experiment were full factorial. The experimenter starts with what appears to be a low-resolution experiment, but when the trials are completed, the practical result will be a higher-resolution experiment. If factors A and B are influential, four treatment conditions have been tested, which makes the experiment full factorial for those factors. The dilemma at the start of experimentation is not knowing which factors are really influential.

Utilization of linear graphs. The first edition of this text placed a fair amount of emphasis on linear graphs and understanding modification of linear graphs to assign factors. The concept of linear graphs and modification of linear graphs has been confusing to many technical people and is not used very much. Table D-3 (two-level OAs) in App. D has the recommended column assignments for all the different situations up to an L32 OA which is much easier to use than linear graphs. Table D-4 (three-level OAs) shows similar arrangements up to an L27 OA. These assignments were derived using the interaction tables to provide the highest resolution, so there is no need to recreate this information each time an experiment is designed. Because of this approach, the use of linear graphs has been excluded from the second edition.

3-4   Recommended Experiment Design Approach Summary

Once a list of factors and levels has been determined, the next major part of the planning phase is to determine the actual experimental test combinations that will evaluate the factors at their various levels. This is a two-part project. First, choose an orthogonal array to use, considering the number of factors, the number of levels of the factors, resolution, and cost (Table D-1 or D-2); second, assign factors to columns in the chosen orthogonal array (Table D-3 or D-4). When factors are assigned to positions in an OA, this automatically dictates all the possible combinations of factors and levels which will be tested. Steps 7 through 9 of the DOE process described in Chap. 2 cover these two tasks.

3-4-1 Choosing orthogonal array examples

The water pump and die-cast piston experiments are discussed in the following sections with respect to which OA should be used and interactions that may be considered. The factors and levels are referenced in Table 2-3.

Water pump experiment. With 7 two-level factors in an L8 OA, there will be no empty columns; in an L16 OA there will be eight columns that have no factors assigned. In the L16 case, one more two-level factor can be evaluated without decreasing the resolution and more factors can be added without increasing the experiment size (resolution will decrease, however). If additional factors are not assigned to the OA, then specific interactions may be identified as interesting to evaluate. With the seven original factors, the recommended choice is to use an L8 OA to maintain a small experiment for the beginning round.

Die-cast piston experiment. With 5 two-level control factors in an L8 OA, there will be two unassigned columns. Two interactions or two additional factors could be evaluated without increasing the size of the experiment. One possible interaction is the water-cooling and air-cooling interaction; both of these factors relate to heat-treating principles of an alloy. The position on a piston is treated as a noise factor and can be assessed by collecting hardness data on several pistons within a trial on the dome and the skirt positions.

3-4-2 Assigning factors to OA examples

The appropriate column assignments for the factors in the water pump and die-cast piston experiments are the pertinent pieces of information relevant to this section.

Water pump experiment. According to Table D-1, 7 two-level factors in a low-resolution situation would use an L8 OA, and, in a medium resolution situation, they would use an L16 OA. The L8 OA is recommended as a beginning experiment to minimize the number of tests, test time, and test cost. Since there are seven factors and seven columns in an L8 OA, references to the factor assignment Table D-3 are not necessary; each factor is assigned to a column. The particular column to which a factor is assigned is not important since all the combinations of column assignments will be statistically equivalent. The trial combinations will be different for each of the different column assignments, but each factor will be evaluated at the first and second levels an equal number of times.

Die-cast piston experiment.   According to Table D-1, 5 two-level control factors in a low-resolution situation, they would use an L8 OA; in a medium-resolution situation, they would use an L16 OA; and in a high-resolution situation, they would use an L32. An L8 OA is recommended as a beginning experiment to minimize the number of tests, test time, and test cost. Factor assignment Table D-3 indicates that the five factors should be assigned to columns 1, 2, 4, 7, and the fifth factor to either column 3, 5, or 6. The noise factor of piston position need not be included in the OA since the hardness of each test piston can be measured in two places. Since this is a resolution 1 experiment, there are several equivalent assignments, even though they are different from that recommended in Table D-3. An equivalent assignment actually used was factors *A, B, C, D,* and *E* (reference Table 2-3) in columns 1, 2, 4, 5, and 6, respectively. This actual assignment will be utilized throughout the remainder of the text.

3-5   Multiple-Level Experiments

This chapter deals with modification of standard two-level arrays to handle factors of three or four levels. Three-level arrays will handle three-level factors without modification, but a mixture of two-, three-, and four-level factors requires special treatment of standard two-level arrays.

3-5-1   Necessity for multiple-level experiments

As was mentioned in the previous chapter, there are occasions when discrete factors will be assigned to an experiment and more than two levels will be required. There are situations involving continuous variables that may make three levels very useful; for example, the steering wheel position in an automobile may be set in the left, center, or right positions. However, it is still recommended for an initial experimental investigation to begin with two levels whenever possible. Once significant factors are identified, multiple levels can be used for estimating nonlinear responses.

3-5-2   Conversion from two to four levels

A two-level array can be converted to contain some four-level columns very simply. The concept for conversion depends upon the concept of degrees of freedom which will be explained fully in Chap. 5. However, at this point it should be intuitive that a four-level factor will require a larger experiment than a two-level factor. Each column in a two-level OA has one piece of information associated with it: the comparison of the average of all of the first-level tests compared to the average of all of the second-level tests. A four-level factor has three pieces of information associated with it: the first- to second-level effect, the second-to third-level effect, and the third- to fourth-level effect. To statistically accommodate a four-level factor in a two-level OA, three two-level columns must be replaced with one four-level column which will provide the same information potential. It is recommended to merge three columns which are mutually interactive, such as columns 1, 2, and 3. Any set of mutually interactive columns may be found by choosing two columns and determining the interaction column from the interaction table in App. B. The merging of mutually interactive columns minimizes confounding of interactions as much as possible and also maintains the orthogonality of the final array.

If columns 1, 2, and 3 are merged in an L8 to form a single four-level column, the eight trials need to have a particular level assigned to maintain the orthogonality of the array. A recommended technique for accomplishing this is as follows. The first three columns of an L8 array appear in Table 3-12. If any two columns are studied, one will notice that there are four combinations possible of levels 1 and 2.

$$1\ 1 \quad 1\ 2 \quad 2\ 1 \quad 2\ 2$$

Level 1 for the four-level factor can correspond to the 1 1 condition, level 2 to the 1 2 condition, level 3 to the 2 1 condition, and level 4 to the 2 2 condition. Using columns 1 and 2 in Table 3-12, the corresponding levels would appear as in Table 3-13. The three merged columns provide the necessary amount of information and any two of those columns provide an orthogonal pattern for the four levels. The four levels need not be in consecutive order with respect to the trial numbers; in this case, the foremost requirement is the balanced number of two trials at each level.

The resultant L8 array modified to include one four-level factor and up to four two-level factors is shown in Table 3-14. The columns have

TABLE 3-12   L8 OA Columns 1, 2, and 3

	Column no.		
Trial no.	1	2	3
1	1	1	1
2	1	1	1
3	1	2	2
4	1	2	2
5	2	1	2
6	2	1	2
7	2	2	1
8	2	2	1

TABLE 3-13   Four-Level Factor Arrangement

Trial no.	Column no. 1	2	3	4-Level factor
1	1	1	1	1
2	1	1	1	1
3	1	2	2	2
4	1	2	2	2
5	2	1	2	3
6	2	1	2	3
7	2	2	1	4
8	2	2	1	4

TABLE 3-14   L8 OA Modified for a Four-Level Factor

Trial no.	Column no. 1	2	3	4	5
1	1	1	1	1	1
2	1	2	2	2	2
3	2	1	1	2	2
4	2	2	2	1	1
5	3	1	2	1	2
6	3	2	1	2	1
7	4	1	2	2	1
8	4	2	1	1	2

been renumbered in this array. Many other combinations of the four-level factor and the remaining two-level columns are possible. Note that it is not possible to include two four-level factors in an L8 array without causing some confounding of portions of the factor effects. If three mutually interactive columns are selected, another set cannot be found that does not have some column in common. More than one four-level factor may be included in an L16 OA; in fact, up to five four-level factors may be utilized.

Appendix B contains several modified OAs having one or two four-level columns. It is recommended to use these modified OAs if the need should arise.

3-5-3   Multiple-level factor interactions

Occasionally, it may be necessary to include an interaction of a two-level factor and a four-level factor. The interaction can obviously be evaluated in an OA, but some columns must be allocated for that interaction. If the four-level factor A was created by merging three mutually interactive columns 1, 2, and 3, a separate two-level factor B could be assigned to a column—say column 2 (original column 4)—in the two-level array. Then, the interaction effect would be estimated in all the columns that interact between original column 4 and original columns 1, 2, and 3, which would be original columns 5, 6, and 7 (new columns 3, 4, and 5). Recall that the estimate of the interaction effect is obtained by adding the effects of columns 5, 6, and 7. The portion of the total effect represented in any column is unpredictable; only the total can be used for effect due to the interaction. Since all the factor and interaction effects are in separate columns with no confounding, this is a resolution 4 experiment. If any one two-level factor is added to the experiment, the resolution would drop to a value of 1 since a factor is confounded with a two-factor interaction.

3-5-4   Three-level factors (dummy treatment)

The dummy treatment method is based on the four-level method when modifying a basic two-level OA. Three mutually interacting columns are merged and the pattern for the four levels is determined. The dummy treatment accommodates three-level factors by using only three of the four possible levels for the factor and the indicated fourth level simply repeats one of the previous three levels. Any one of the three levels for the factor can be repeated as shown in Table 3-15. Any one of the three levels can be repeated and maintain orthogonality, so whichever is easiest, cheapest, or makes more sense should be repeated. The factor and interaction locations are equivalent to a four-level factor situation.

TABLE 3-15   L8 OA Dummy Treatment

Trial no.	4 Levels Merged columns (1 2 3)	3 Level options Merged columns (1 2 3)		
1	1	1 or 1 or 1		
2	1	1	1	1
3	2	2	2	2
4	2	2	2	2
5	3	3	3	3
6	3	3	3	3
7	4	1'	2'	3'
8	4	1'	2'	3'

NOTE: ' indicates the repeated level; only one of the options may be used in one experiment.68

3-6   Special Designs

Some situations occur in typical product or process development projects which are not accommodated by standard two- or three-level OAs. The following sections describe three special OA arrangements.

3-6-1   Nested experiments

This situation arises when experimenting with discrete (noncontinuous) factors. The use of a discrete factor may cause each half of a two-level OA to have entirely different kinds of test conditions. One example proposed two kinds of retention methods, factor A, for a plug inserted into the end of a valve. One retention method was to simply press-fit the plug into place and the other was to heat-stake the plug after press-fit assembly. The factors that apply to the heat-staking process are not at all relevant when only press fitting the plug into position. If the retention method is assigned to the first column of an OA, then the upper half of the experiment could be related to heat staking and the lower half to not heat staking. If a heat-staking factor, such as the amount of heat-staking time, factor B, is assigned to the second column, levels 1 and 2 in the lower half of the experiment are meaningless with respect to not heat staking (see Table 3-16). This condition requires the use of a nested experiment to accommodate the discrete factor(s). The heat-staking time factor is nested within the retention method factor.

This approach is recommended if there are relatively few nested factors and many common factors to be evaluated. If there are many nested and only a few common factors, then two separate experiments should be done to optimize the two nesting factor options. After

TABLE 3-16   Nested Factor Column Assignment and OA Modification

	Factors		
	A	B	
	Column no.		
Trial no.	1	2	3
1	1	1	~~1~~
2	1	1	~~1~~
3	1	2	~~2~~
4	1	2	~~2~~
5	2	1	~~2~~
6	2	1	~~2~~
7	2	2	~~1~~
8	2	2	~~1~~

each method is optimized, then the two options may be compared against each other to determine the best overall method.

The nested experiment requires a special modification to accommodate the discrete factor(s). The first step is to identify the nesting factor(s) in the experiment. These are the factors whose levels are varied over the entire experiment and cause the experiment to be split into two major portions. The retention method is a nesting factor and can be assigned to column 1, as an example. Second, the nested factors are then assigned to columns in the OA, column 2, as an example.

Trials 1 to 4 in column 1 relate to heat staking and trials 5 through 8 relate to press fitting. Trials 1 to 4 in column 2 relate to low and high time of heat staking and trials 5 to 8 relate to low and high values for experimental error estimation. Since there are two pieces of information effectively available in column 2, the third column is omitted to allow the necessary information to be available as shown in Table 3-16. The third column is the mutually interactive column of the nesting and nested factors. This technique is similar to the four-level factor merging concept previously discussed.

Other nested factors can be joined to nesting factor A by using the unassigned columns and eliminating interacting columns. Factor C, such as temperature of heat staking, could be nested in factor A. Factor C would have to be assigned to a separate column; consequently, the interaction column for $A \times C$ would be omitted to provide a piece of information for experimental error in the factor C column. Table 3-17 contains a potential list of factors for this experiment. Factor E applies only when the parts are not heat-staked, so a more compact OA arrangement may be used. Since the lower half of column 2 was unassigned (error estimate), two factors can be assigned to this column. The upper half of column 2 applies to factor B and the lower half applies to factor E. Each factor will be evaluated in the half of the experiment to which it pertains. Table 3-18 shows the complete factor assignment. Nesting factor A and common factor D are varied over the entire experiment, whereas nested factors B, C, and E are effectively varied in only one-half of the experiment.

TABLE 3-17   Heat-Staking Factors

Factors	Level 1	Level 2
A retention method	Heat-staked	Not heat-staked
B stake time	Low	High
C stake temperature	Low	High
D press-fit amount	Low	High
E press-fit depth	Flush	Below

TABLE 3-18   Nested Factor OA Column Assignment

	Factors				
	A	B,E	C,e	D	A × D
			Column no.		
Trial no.	1	2	3	4	5
1	1	1	1	1	1
2	1	1	2	2	2
3	1	2	1	2	2
4	1	2	2	1	1
5	2	1	1	1	2
6	2	1	2	2	1
7	2	2	1	2	1
8	2	2	2	1	2

(Column 2: rows 1–4 grouped as B, rows 5–8 grouped as E. Column 3: rows 1–4 grouped as C, rows 5–8 grouped as e.)

Note: e = official experimental error estimate.

3-6-2   Combined factors method

Sometimes the factors selected for evaluation include a mixture of two-level and three-level factors. This method is used when there are mostly three-level factors. A two-level array can be used in conjunction with the dummy treatment method for each three-level factor treated in this fashion. A three-level array may also be used and the two-level factors be dummy-treated, which does not fully utilize the size of the OA. If the factor list is at the limit of the number of columns available in an OA and the next-larger OA is not desirable from a cost viewpoint, rather than eliminate factors to fit the OA, it is better to include them in the experiment in some form.

The combined factors method treats a pair of two-level factors as a three-level factor. However, the estimate of any interaction between those factors must be sacrificed. The four possible combinations of two factors at two levels are, for example,

$$A_1B_1 \quad A_2B_1 \quad A_1B_2 \quad A_2B_2$$

If the A_1B_1 condition, A_2B_2 condition, and one of the remaining two conditions are selected, then two comparisons can be made to estimate the factor A and B effects. Assuming these test conditions:

Combined factor level	1	2	3
Factor A/B levels	A_1B_1	A_2B_1	A_2B_2

Comparing combined factor level 1 to level 2 provides an estimate of factor A's effect (factor B is held constant at B_1) and level 2 compared to level 3 provides an estimate of factor B's effect (factor A is held con-

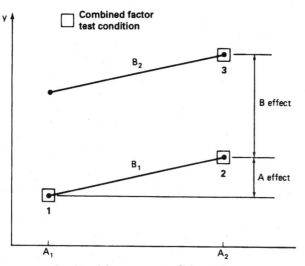

Figure 3-6 Combined factors test conditions.

stant at A_2). Figure 3-6 shows how this test arrangement can esti-mate the factor effects when these particular treatment conditions are selected. If there happens to be an interaction between factors A and B, then one of the factor effects will be overestimated and one will be underestimated. However, no knowledge of which factor is overestimated and which is underestimated will be available; it will not be known whether the nonparallel lines for an interaction would be converging on the left or right side of the graph.

These three test conditions can be assigned to one column in a three-level OA as a combined factor AB, as shown in Table 3-19. Because the A_1B_2 condition is not tested, the orthogonality of factors A and B is lost.

3-6-3 Idle column method

The idle column method can be used to fit several three-level factors into a two-level OA with many two-level factors. This method also partially destroys orthogonality but it eliminates the need for dummy treatment and the subsequent deletion of a column for each three-level factor. For two or more three-level factors, the idle column method tends to reduce the size of the experiment but at the sacrifice of orthogonality of the three-level factors.

The idle column should be assigned to column 1 of a two-level OA and becomes the level indicator for the three-level factors when they are assigned to a two-level column. Three-level factor A is assigned to column 2 and the interaction column 3 is eliminated. Three-level fac-

TABLE 3-19   Combined Factors OA Column Assignment

	Factors			
	AB	C	D	E
	Column no.			
Trial no.	1	2	3	4
1	1	1	1	1
2	1	2	2	2
3	1	3	3	3
4	2	1	2	3
5	2	2	3	1
6	2	3	1	2
7	3	1	3	2
8	3	2	1	3
9	3	3	2	1

tor B is assigned to column 4 and the interaction column 5 is eliminated. The idle column is always the common column in the mutually interactive groups where three-level factors are assigned.

The levels for the three-level factors are dictated by the levels indicated in the idle column. When the idle column indicates level 1, then levels 1 and 2 are assigned for the three-level factor. When the idle column indicates level 2, then levels 2 and 3 are assigned for the three-level factor. The OA is set up for factor A as shown in Table 3-20 for an L8 OA. Levels 2 and 3 are substituted for 1 and 2, respectively, when the idle column indicates level 2. The interacting column must be eliminated from the OA; in this case, column 3 was removed.

If 2 three-level factors (A and B) and 2 two-level factors (C and D) were assigned to an L8 OA using column 1 as the idle column, the

TABLE 3-20   L8 OA Idle Column Assignment

	Factors		
	Idle	A	
	Column no.		
Trial no.	1	2	3
1	1	1	1
2	1	1	1
3	1	2	2
4	1	2	2
5	2	1̶2	2
6	2	1̶2	2
7	2	2̶3	1
8	2	2̶3	1

TABLE 3-21   Idle Column Factor Assignment

			Factors		
	Idle	A	B	C	D
			Column no.		
Trial no.	1	2	3	4	5
1	1	1	1	1	1
2	1	1	2	2	2
3	1	2	1	2	2
4	1	2	2	1	1
5	2	2	2	1	2
6	2	2	3	2	1
7	2	3	2	2	1
8	2	3	3	1	2

assignment would appear as in Table 3-21. Original columns 3 and 5 were removed from the OA.

3-7 Summary of Multiple-Level Methods

Four different methods of accommodating multiple levels have been reviewed in the text:

1. *Merging columns.*   A four-level factor can be fit into a two-level OA (see Secs. 3-5-2 and 3-5-3).

2. *Dummy treatment.*   A three-level factor can be fit into a two-level OA (see Sec. 3-5-4).

3. *Combination method.*   Two-level factors can be fit into a three-level OA (see Sec. 3-6-2).

4. *Idle column method.*   Many three-level factors can be fit into a two-level OA (see Sec. 3-6-3).

The first two methods do not disturb the orthogonality of the entire experiment. The last two methods cause a loss of orthogonality among the factors treated by those methods and subsequently the loss of an accurate estimate of the independent factorial effects. The advantage of the last two methods is a smaller experiment when test costs are very high.

3-8 Component Identification Design

This method is an adaptation of the use of two-level OAs when the reassembly and retest of a product are possible. A test that is destructive in nature is very difficult if not impossible to do in this manner. This method is primarily a cause-detection type of experiment.

A component (or subassembly) identification experiment is possible if a product is tested and some units are found to be relatively good performers and others poor performers. The experiment can be completed with as little as one good unit and one bad unit but with some limitations which will be discussed later.

The component identification method begins by determining the number of major components or subassemblies within a total assembly that might affect the performance characteristic of interest. The number of selected components dictates the size of the OA; again, at least one column in an OA will have to be allocated to each component to be evaluated.

Let's assume that a certain percentage of automatic transmission hydraulic pump assemblies are failing the pressure and flow test at the final test for pumps. If seven major components are identified, then an L8 OA can be the basis for the experiment. The next step would be to collect eight pump assemblies that are poor pressure and/or flow performers and eight pumps that are good performers. It may take some time to collect the eight poor assemblies from a production line unless the problem is severe. The hypothesis behind this experiment is that there are one or more reasons which consistently cause the differences between the good and poor performers.

The pumps are disassembled into groups of common components with all the good components segregated from the poor components. For instance, two groups of eight pump drive gears each should exist when the disassembly is complete: eight gears from the good units and eight gears from the bad units. The pump assemblies are rebuilt into 16 complete assemblies by using the L8 OA as shown in Table 3-22. In this OA, level 1 can symbolize a component from a poor performing unit and level 2 can symbolize a component from a good performing unit (or vice versa). Two pump assemblies can be built in the combination indicated in each trial. The components should be randomly selected from the poor group of components when a level 1 is shown in the OA and randomly selected from the good group when a level 2 is shown.

The data should indicate a propensity to fall into a pattern that will match the column pattern for the component(s) contributing to the problem. Hopefully, experimental error within a trial will be small, which indicates that disassembly and reassembly do not affect the pump performance. If the experimental error is large, then excess measurement system error and/or assembly sensitivity is indicated.

If only four good units and four poor units are used in this experiment, then the effect of disassembly and reassembly can be evaluated by completely disassembling, reassembling, and retesting the pumps.

TABLE 3-22   Component Identification Experiment

	Components								
	A	B	C	D	E	F	G		
	Column no.							Pump flow	
Trial no.	1	2	3	4	5	6	7	y_1	y_2
1	1	1	1	1	1	1	1	*	*
2	1	1	1	2	2	2	2	*	*
3	1	2	2	1	1	2	2	*	*
4	1	2	2	2	2	1	1	*	*
5	2	1	2	1	2	1	2	*	*
6	2	1	2	2	1	2	1	*	*
7	2	2	1	1	2	2	1	*	*
8	2	2	1	2	1	1	2	*	*

The two tests for each trial should be very consistent unless reassembly affects performance or measurement system error exists since the identical components are retested.

This method can be used with as little as one good unit and one poor unit by using the parts to build the particular combination specified for each trial. All eight combinations described by the OA would be obtained by a complete teardown and rebuild between trials. If this assembly were torn down, completely and identically reassembled, and retested, then an estimate of experimental error would be obtained.

An example of using the latter technique involves a problem with manual transmission production. A squealing noise in third range was noted in some gearboxes on the final test stands. One noisy and one quiet gearbox were selected from production. Several components related to third range were chosen as items of interest in the investigation. Six items were selected which allowed the use of an L8 OA with one column available for the rest of the transmission assembly from each noisy and quiet gearbox. The factors for the experiment are listed in Table 3-23. The assignment of factors to an L8 OA and the subsequent test results are shown in Table 3-24. The entire experiment was completed in six hours, since the response was an immediate performance characteristic rather than a long durability characteristic.

In this experiment, no method was used to quantify the noise level; only a notation was made as to whether the noise was present or not. The pattern of the noise response matched exactly the pattern of levels for column 7, the third-range gear. Some test engineers were skeptical of the conclusions that the gear was the only problem and ran several other combinations of parts in an attempt to support their theory. However, the noise remained associated with the third-range gear only.

TABLE 3-23   Noisy Transmission
Component Identification Experiment

Factors	Level 1	Level 2
A sleeve	Noisy	Quiet
B fork	"	"
C transmission assembly	"	"
D input shaft	"	"
E bearing	"	"
F blocker	"	"
G gear	"	"

TABLE 3-24   Transmission Third-Range Noise Experiment

	Factors							
	A	B	C	D	E	F	G	
	Column no.							
Trial no.	1	2	3	4	5	6	7	Noise data
1	1	1	1	1	1	1	1	Squeal
2	1	1	1	2	2	2	2	Quiet
3	1	2	2	1	1	2	2	Q
4	1	2	2	2	2	1	1	S
5	2	1	2	1	2	1	2	Q
6	2	1	2	2	1	2	1	S
7	2	2	1	1	2	2	1	S
8	2	2	1	2	1	1	2	Q

The noisy and quiet gears were then compared dimensionally to isolate any physical differences that existed between the parts. The component identification method indicates which components cause the problem, but not the specific root cause in terms of product parameters. Once a list of differences is established, an experiment should be designed to investigate those factors at the levels that were found in the noisy and quiet gears. In this situation, the only significant difference between the quiet and noisy third-range gears was determined to be the shape of the involute profile. Experiments were then conducted to determine the practical limits for deviation from a true involute profile.

In this case study, the cause of noise was associated with one particular component. Had the noise been associated with more than one component or an interaction between components, the noise level would have more variation than just the two values of noisy and quiet. If this were the situation, then a more sophisticated measurement system for noise would probably have to be used to obtain meaningful results. Also, a more sophisticated analysis method would be required to interpret the results effectively; see Chap. 5 for analysis methods.

A test engineer in the noise lab commented when the test was complete, "We've learned more in six hours using this method than we have in months on previous problems using other methods." A sparse amount of data that is collected strategically is worth more than a large amount of data with a poor collection strategy.

3-9 Summary

This chapter covered some of the basic weaknesses of typical test strategies and offered some more efficient and powerful alternative strategies in the form of orthogonal arrays. Also covered were the basic steps necessary to designing OA experiments. These fundamentals are necessary for utilizing OAs in some more complex situations such as multiple-level experiments. Discussed in this chapter were the methods for modifying a standard two-level OA to handle a three- or four-level factor and some special experimental design situations typically faced by the engineer or scientist. The nested factors section illustrated the method of handling a situation where a factor causes two completely different treatments in halves of an experiment. Methods of accommodating two-level factors in three-level OAs and three-level factors in two-level OAs were discussed. A summary of the possible OA modifications was covered in Sec. 3-7. An interesting application of OAs is the component identification method when reassembly and retest of a device is possible, as covered in Sec. 3-8.

This chapter completes the planning phase of the DOE process. The next major phase is conducting the experiment, which is covered in Chap. 4.

Problems

3-1 What orthogonal arrays are appropriate for experimenting with 4 two-level factors?

3-2 What resolution is provided with 5 two-level factors using an L16 orthogonal array?

3-3 What orthogonal array is probably the best choice for evaluating 8 two-level factors? Why?

3-4 What resolution is provided with 2 three-level factors in an L9 orthogonal array?

3-5 What resolution is provided with 4 three-level factors in an L27 orthogonal array?

3-6 To what columns should 4 two-level factors be assigned in an L8 orthogonal array?

3-7 To what columns should 4 two-level factors be assigned in an L16 orthogonal array?

3-8 To what columns should two-level factors A, B, C, D, and E and interactions A × C and A × D be assigned in an orthogonal array with the lowest resolution?

3-9 To what columns should 7 two-level factors be assigned in what orthogonal array to obtain a resolution 2 design?

3-10 What orthogonal arrays are appropriate for the evaluation of 1 four-level factor and 5 two-level factors?

3-11 To what columns should 4 three-level factors be assigned in an L27 orthogonal array?

3-12 To what columns should 6 three-level factors be assigned in an L27 orthogonal array?

3-13 What is unique about columns 1, 2, and 3 or columns 1, 6, and 7 in a two-level orthogonal array?

3-14 What column in a two-level orthogonal array would estimate the interaction effect A × B, if factor A were assigned to column 1 and factor B to column 5 (not necessarily the recommended column assignment)?

3-15 What column in a two-level orthogonal array would estimate the interaction effect A × B × C, if factor A were assigned to column 1, factor B to column 4, and factor C to column 6 (not necessarily the recommended column assignment)?

3-16 Which factor is usually assigned to a column in an orthogonal array first, the nesting or nested factor?

3-17 If nesting factor A is assigned to column 1 in a two-level orthogonal array and nested factor C is assigned to column 4, which column must be eliminated from the array?

3-18 Do any columns need to be eliminated using the combined factors method to include 2 two-level factors in an L9 orthogonal array?

3-19 What is the main advantage of using the idle column method when there are nearly equal numbers of two-level and three-level factors to be evaluated?

3-20 What is the main limitation of the component identification experimental design approach?

Conducting Tests

4-1 Testing Logistics

Once the chosen factors are assigned to a particular column of the selected OA, the test combinations are set and physical preparation for performing the tests can begin. Main items to consider in the logistics of testing are

1. A system to describe the various test combinations in operational terms (trial description and data sheets)

2. A plan to acquire materials needed for the various test combinations

3. A plan to acquire access to test and production equipment

4. A plan to acquire access to measuring equipment

5. A definition of roles and responsibilities of those involved in conducting the experimental trials

6. A plan for identifying the results of the experimental trials (tagging parts with trial and repetition numbers, for example)

4-1-1 Description of test combinations

Factors are assigned to columns in a given OA; trial test conditions, however, are dictated by the rows. Referring once again to the water pump leak example, each of the seven factors was assigned to a column in an L8 OA. The OA and assignment would appear as in Table 4-1. One can observe that trial 6 requires the test conditions of $A_2B_1C_2D_2E_1F_2G_1$. Trial 3 requires test conditions of $A_1B_2C_2D_1E_1F_2G_2$. This, obviously, is not a convenient way to describe the eight different test combinations. Trial description and data sheets should be pre-

TABLE 4-1   Water Pump Experiment

	A	B	C	D	E	F	G
				Factors			
				Column no.			
Trial no.	1	2	3	4	5	6	7
1	1	1	1	1	1	1	1
2	1	1	1	2	2	2	2
3	1	2	2	1	1	2	2
4	1	2	2	2	2	1	1
5	2	1	2	1	2	1	2
6	2	1	2	2	1	2	1
7	2	2	1	1	2	2	1
8	2	2	1	2	1	1	2

pared which describe these eight combinations in operational terms. For example, referring to the factor levels for the water pump experiment in Table 2-3, trial 6 is operationally described by this assembly:

Factor	Trial 6 level
Cover design	new
Gasket design	production
Front bolt torque	USL
Sealant	yes
Pump finish	rough
Back bolt torque	USL
Torque sequence	F-B

A separate sheet should be constructed for each of the eight trials to avoid any confusion as to what combination of factors and levels will be used in any given trial.

The interaction conditions cannot be controlled when conducting a test because they are dependent upon the main factor levels. Only the analysis is concerned with any interaction columns. Therefore, it is recommended that test sheets be made up which show only the main factor levels required for each trial. This will minimize mistakes in conducting the experiment which may inadvertently destroy the orthogonality.

4-2 Statistical Aspects of Conducting Tests

There are two main statistical considerations involved in conducting experiments: (1) the statistically valid sample size and (2) the randomization strategy used to determine test order.

4-2-1 Sample size determination

The type of quality characteristic under evaluation, variable or attribute, has a tremendous effect on the sample size required in the experiment. In general, variable data will require considerably fewer tests compared to attribute data to reach the same statistical confidence that factors make a difference in the quality characteristic(s).

Variable data. Variable data is a continuous form, which means that an infinite number of values can occur anywhere between very low values and very high values. The discrimination between any two experimental results depends only on the precision of the measurement system. Examples of variable data are temperature, pressure, flow rate, efficiency, power, weight, length, velocity, volume, and time. The temperature produced by a chemical reaction might be a performance characteristic of interest (instant cold packs) which could technically assume any value between $-460°F$ and an extremely large positive value of temperature.

As was mentioned in Sec. 3-1, there are two textbooks that provide excellent explanations for the determination of the statistically valid sample size. However, from a very practical viewpoint, a minimum of one test result for each trial is required to maintain the sample size balance (orthogonality) of the experiment. (If the test results are unbalanced, then the experiment requires a special analysis not covered in this text.)

More than one test per trial can be used, which increases the sensitivity of the experiment to detect small changes in averages of populations. An economic consideration also can be made at this time. If tests are very expensive, then one or few tests per trial can be used with decreased sensitivity; larger differences in average results will be detected. If tests are inexpensive, then many repetitions per trial can be used with increased sensitivity; smaller differences in average results will be detected.

An L8 OA with one test per trial (four tests versus four tests) makes the experimenter 90 percent sure (confident) of detecting a change in average of approximately 2.4 standard deviations (S). This statement is based on assumptions of a student's t distribution, equal sample size of the two levels, and equal variance within the two levels of factors. An L8 OA with two repetitions or an L16 OA with one repetition per trial (8 versus 8) makes the experimenter 90 percent sure of detecting a change of approximately 1.5 standard deviations. An L16 OA with two repetitions per trial provides 90 percent confidence in detecting a change in average of about 1.0 standard deviation. This is a fairly sensitive experiment, and sample sizes larger than this do not add much to the sensitivity. Experiments as small as an L4 OA should be avoided for the main array; the sensitivity with only one

TABLE 4-2   Sensitivity of Experiments: L8 and L16 OAs

No. of repetitions	Total sample size	Confidence level*		
		90%	95%	99%
L8				
1	8	2.39	3.10	4.84
2	16	1.55	1.95	2.80
3	24	1.24	1.55	2.17
4	32	1.06	1.32	1.84
5	40	.95	1.17	1.63
6	48	.86	1.07	1.47
L16				
1	16	1.55	1.95	2.80
2	32	1.06	1.32	1.84
3	48	.86	1.07	1.47
4	64	.74	.92	1.26
5	80	.66	.82	1.12
6	96	.60	.74	1.02

*Confidence that a difference of the multiples of standard deviation will be detected by the experiment.

test per trial is about 4.8 standard deviations at 90 percent confidence. If more repetitions are added for sensitivity, then an L8 OA should be used which will evaluate interactions as well as provide the increased sensitivity with the same number of total tests.

Table 4-2 shows the sensitivity of various numbers of repetitions for an L8 and L16 OA. The table shows that as the number of repetitions (total sample size) increases there is a diminishing return in the increase of sensitivity of the experiment. Experiments that can detect a shift in average of 1.0 standard deviation are fairly sensitive and experimenters are typically in search of larger effects than this. One can see that even at a high confidence level of 95 percent, no more than six repetitions are necessary to achieve a 1 standard deviation sensitivity. Therefore, as a guideline, no more than six repetitions should be used in an experiment, regardless of the size of the OA. Also, an L4 OA should be used only for applications discussed in Chap. 7, "Parameter Design."

Attribute data.   Attribute data, on the other hand, is a discontinuous form, which means the experimental results can only be discrete values such as good and bad, or off and on, or 0 and 1. A single-pole, single-throw switch can assume either the off position or the on position; there is no such thing as half-off, quarter-off, or quarter-on position. In a production process, a defect of a product is either present or not present. Numerically, this situation can be represented by assigning a value of 1 to a part with a defect and a 0 to a part without a defect (or vice versa for analytical purposes).

Two-class attribute data provides much less discrimination than variable data. When a part is classified as bad, no measure of how bad is provided. When a part is classified as good, no measure of how good is provided. Because of this reduced discrimination, many pieces of attribute data are required to provide the equivalent information of one piece of variable data. Variable data should be used whenever possible due to the greatly reduced sample size.

Two-class attribute data can be of two types. One kind is a situation where the total number of tests, the number of good results, and the number of bad results are known. For example, a certain number of engine cylinders are ignited during a test run and the number of occurrences of normal combustion and the number of occurrences of detonation recorded. Another kind of attribute data is where the total number of tests is known and only the number of one class of occurrence may be known. For example, an experiment to grow apples would recognize the number of apples on a tree. The number of "nonapples" would not or could not be known.

To improve the discrimination power of attribute data, more classes may be used. For instance, if parts were rated on the severity of a defect present, then several classes may be used:

Class no.	Description
1	No defect
2	Mild defect
3	Moderate defect
4	Severe defect

As the number of classes increases, the data becomes semivariable in nature, which adds power of discrimination. In this instance, the class number has some engineering or scientific meaning; the higher the class number, the more severe the defect. In some classified data, the frequency of occurrence in a particular class may not have such meaning—the number of deaths due to heart attack by city block, for instance. Here the block number does not necessarily indicate a more or less severe condition.

Multiple-class (more than two classes) data has other advantages over two-class data which concerns the ability to detect a shift in average or to detect an increase or decrease in variability. The greater the number of classes, the more a frequency distribution by class will resemble a histogram and provide a clue to the shape of the distribution. With only two classes, the shape of the distribution is barely hint-ed. An improvement in uniformity will result in an increase in frequency of one or more adjacent classes with an accompanying decrease in the frequency of other classes when multiple classes are used.

A guideline for the sample size for the type of data where occurrences and nonoccurrences are known is that the class with the least frequency should have a count of at least 20. For instance, in an experiment to assess defective parts, the total number of defectives should be at least 20. If the past performance has been 10 percent defective, then the total sample size of the experiment should be 200 to produce an expected 20 defectives. These defectives are expected to be spread over all the trials to provide information as to which factors were influencing the quantity of defectives. Table 4-3 summarizes the effect of decreasing defective rate on increasing sample size for an L8 OA. When the attribute data is of the type where only the occurrences may be known, the expected number of occurrences should be from two to five per trial.

In the situation of having a very low failure rate, a very large sample size is required to expect 20 failures in the experiment. For example, if the failure rate during customer use is 0.25 percent, then the sample size for the experiment becomes 8000. Rather than use this large a sample size, which may be expensive, it is recommended to intensify the test conditions used in the experimental trials to increase the failure rate and correspondingly decrease the required sample size. The failure mode(s) must remain the same, but the product or process changes that reduce the failure rate on the intensified test should do the same in customer usage. Generally, intensified tests should be used when a reduction in test time and/or sample size is needed (the failure mode must remain the same, which may limit the amount of intensification).

4-2-2 Randomization strategies

Some decisions need to be made concerning the order of testing the various trials. The order of performing the tests of the various trials should include some form of randomization. The randomized trial

TABLE 4-3 Attribute Data Sample Size for L8 OA

Expected defective percentage	Minimum repetitions per trial	Total sample size	Total expected occurrences
20	13	104	21
10	25	200	20
5	50	400	20
1	250	2000	20
0.5	500	4000	20
0.1	2500	20000	20

order protects the experimenter from any unknown and uncontrolled factors that may vary during the entire experiment and which may influence the results. Presuming that the random order does not match any of the patterns described by the columns, such as all odd-numbered trials are tested and then all even-numbered trials (matches the pattern of column 4 of an L8; all 1s and then all 2s), any unknown or uncontrolled factor effects will be spread evenly over the entire experiment. This will prevent a bias in the interpretation of which factors and interactions cause a change in the average of the quality characteristic(s) of interest.

Randomization can take many forms, but the three most used approaches will be discussed.

1. Complete

2. Simple repetition

3. Complete within blocks

Complete randomization. Complete randomization means any test has an equal chance of being selected for the first test. Of the remaining tests, each one has an equal opportunity of being selected for the next test, and so on. In order to determine the complete testing sequence, random number tables, a random number generator, or simply drawing numbers from a hat will suffice. However, even complete randomization may have a strategy applied to it. For instance, several repetitions of each trial may be necessary, so each trial should be randomly selected until all trials have one test completed. Then each trial is randomly selected in a different order until all trials have two tests completed. The experiment will progress on a sequential basis with the opportunity for analysis at the end of each round of repetitions. This method is used when a change of test setup is very easy or inexpensive and the resulting test pattern is shown in Table 4-4.

Simple repetition. Simple repetition means that any trial has an equal opportunity of being selected for the first test, but once that trial is selected, all the repetitions are tested for that trial. This method is used if test setups are very difficult or expensive to change. The simple repetition method is shown in Table 4-5.

Complete randomization within blocks. Complete randomization within blocks is used where one factor may be very difficult or expensive to change the test setup for, but others are very easy. If factor A were difficult to change, then the experiment could be completed in two halves or blocks. All A_1 trials could be randomly selected to be conducted and then all A_2 trials could be randomly selected. This arrangement is shown in Table 4-6.

TABLE 4-4   Complete Randomization (Sequentially)

			Factors					
	A	B	C	D	E	F	G	
				Column no.				
Trial no.	1	2	3	4	5	6	7	y data
1	1	1	1	1	1	1	1	* * *
2	1	1	1	2	2	2	2	* * *
3	1	2	2	1	1	2	2	* * *
4	1	2	2	2	2	1	1	* * *
5	2	1	2	1	2	1	2	* * *
6	2	1	2	2	1	2	1	* * *
7	2	2	1	1	2	2	1	* * *
8	2	2	1	2	1	1	2	* * *

Randomized within first round ——
Randomized within second round ——
Randomized within third round ——

TABLE 4-5   Simple Repetition

			Factors					
	A	B	C	D	E	F	G	
				Column no.				
Trial no.	1	2	3	4	5	6	7	y data
1	1	1	1	1	1	1	1	* * *
2	1	1	1	2	2	2	2	* * *
3	1	2	2	1	1	2	2	* * *
4	1	2	2	2	2	1	1	* * *
5	2	1	2	1	2	1	2	* * *
6	2	1	2	2	1	2	1	* * *
7	2	2	1	1	2	2	1	* * *
8	2	2	1	2	1	1	2	* * *

Random order of trials; do all repetitions for a trial ——

Effects of randomization.   Randomization can be detrimental in some experimental situations. If some sources of variation are known, such as having to use two bags of plastic pellets for an injection molding experiment or two automobiles and drivers for a manual transmission shiftability experiment, then these sources of variation can be treated as factors and assigned to columns. This method of controlling known sources of variation is called *blocking*. If the experimenter chooses to randomize the experimental trials with respect to these known sources of variation, then the experimental error will be inflated if these sources really do cause a difference in product response. If the sources are blocked and do have an effect on response, then this effect can be estimated and experimental error will not be unduly

TABLE 4-6   Complete Randomization within Blocks

				Factors				
	A	B	C	D	E	F	G	
				Column no.				
Trial no.	1	2	3	4	5	6	7	y data
1	1	1	1	1	1	1	1	* * *
2	1	1	1	2	2	2	2	* * *
3	1	2	2	1	1	2	2	* * *
4	1	2	2	2	2	1	1	* * *
5	2	1	2	1	2	1	2	* * *
6	2	1	2	2	1	2	1	* * *
7	2	2	1	1	2	2	1	* * *
8	2	2	1	2	1	1	2	* * *

Random order of trials in block 1
Random order of trials in block 2

inflated, which will protect the statistical power of the experiment. A guideline is to block (treat as another factor) known sources of variation in an experiment. Taguchi has developed a different approach to handling this situation, which will be discussed in Chap. 7 concerning parameter design.

The different methods of randomization affect experimental error in different ways. Complete randomization allows a longer time to transpire between repetitions in a given trial compared to simple repetition. Because of this, unknown and uncontrolled factors that may be varying during an experiment may make the repetition-to-repetition variation larger with complete randomization compared to simple repetition. Simple repetition, because of the generally longer times transpiring between trials, will show larger trial-to-trial variation compared to complete randomization. Increased trial-to-trial variation with decreased repetition variation will tend to make factors appear statistically significant in the analysis phase when, in fact, they are not. This situation will typically occur on factors that appear to make relatively small differences in the average result, because factors that make a substantial change in the average response will tend to show up statistically significant regardless of the randomization strategy. However, because of these considerations, complete randomization is recommended whenever possible.

4-3   Characteristics of Good and Bad Data Sets

Good and bad data sets can be characterized in several ways:

Good data set	Bad data set
All trials and repetitions are complete; balanced data	Missing trials and/or repetitions (Sec. 4-3-1)
Consistent results within a trial; low variation within a trial (Sec. 4-3-4)	Inconsistent results within a trial; high variation within a trial (Sec. 4-3-2)
Large differences in results from trial to trial (Sec. 4-3-4)	Small differences in results from trial to trial (Sec. 4-3-3)

Simply by observing the resultant data, especially with respect to an OA structure, the experimenter can predict the type of information that may be generated from the analysis. Good data sets will typically generate positive information about which factors do make a difference in the quality characteristic of interest. Bad data sets typically will generate negative information about which factors do not make a difference in the quality characteristic of interest. The comments concerning bad data sets will be addressed first.

4-3-1 Missing trials and/or repetitions

Missing trials and/or repetitions should be rerun, if the situation allows. This provides a balanced data set which is the key element of maintaining the experimental orthogonality. However, this is still no guarantee of getting the result(s) that would have been originally obtained. If anything has changed since the original experiment was conducted, it may influence the tests conducted at later time.

Missing repetitions. If the missing repetitions cannot be rerun, then there are several options for the experimenter to proceed to the analysis phase:

1. Analyze the data as an unbalanced data set, which can be accomplished with some difficulty using special statistical analysis computer software packages.

2. Consider the number of repetitions to be equal to the trial having the least number of data points to obtain a balanced data set.

3. Calculate an estimated "replacement" repetition(s) based on the average of the rest of the data within the trial; this is the experimenter's best guess at what the data should be.

From a practical point of view, approaches 2 and 3 should be tried and immediately analyzed. If the two approaches agree very closely, then rerunning the missing data points may prove to be of no advantage. The last two approaches should be used and compared, especially when tests are expensive.

Missing trials. This is a much bigger problem in that there is no information to estimate what the missing trial results should be. The experimenter has these options in order to proceed to the analysis phase:

1. Analyze the data as an unbalanced data set which can be accomplished with some difficulty using special statistical analysis computer software packages.
2. Calculate an estimated "replacement" trial and repetition(s) based on the grand average of the all of the data within the whole experiment; this is the experimenter's best guess at what the data should be.

In this case, it is preferred to rerun the missing trial and repetitions. A larger portion of the experiment is being "estimated" and, therefore, will more significantly bias the interpretation of the final results, causing incorrect conclusions on the part of the experimenter.

4-3-2   Inconsistent results within a trial

Lack of repeatability of repetitions within a trial is caused by two possible items. One, the measurement system may not be repeatable, and, two, the wrong factors have been selected for the experiment. If the measurement system is known or proven to be repeatable, then there are other sources of variation that cause the inconsistency within a trial. The important factors that are left out of the experimental evaluation are changing during the course of conducting the test and cause substantial test-to-test differences in results.

4-3-3   Small differences from trial to trial

Small differences in results from trial to trial are an indication of one of two possibilities. One, important factors may have inadvertently been left out of the experimental evaluation, and, two, the difference between the levels of the chosen factors may be too small to cause any substantial difference.

4-3-4   Good data sets

Good data sets exhibit consistency within a trial and relatively large difference from trial to trial. Both of these conditions are strong indicators that the important factors have been chosen and the levels for those factors represent meaningful changes.

If an experiment is conducted on a product or process where there is a history of the problem being investigated, then the variation in the experimental data should cover at least 75 percent of the variation seen historically. The range in the experiment from good to bad results should be at least 75 percent of the range from good to bad results in recent production data. The reason is to have some indication that the correct factors and levels were included in the experiment.

If the range of variation is too small, then the indication is that the factor which really causes variation has been left out of the experiment or the levels chosen for that factor in the experiment have been selected too closely. The experimenter may not be able to control or reduce variation in production sufficiently with the chosen factors and levels.

If the range of variation is greater than 75 percent of the past, then this still does not completely ensure that the truly influential factors have been included. The factor(s) which really cause variation, but are possibly not included in the experiment, could have been varying unknowingly to provide the high to low values of test data. However, if it turns out the correct factors and levels were chosen, then the experimenter knows that the variation in the production situation can be controlled or reduced substantially.

4-4   Example Experiments

The following examples cover the important issues discussed in this chapter. The last example discusses the advantage of using variable data over attribute data in reducing sample size and improving product performance.

4-4-1   Water pump experiment

An example of one of the trial data sheets for factors and levels, in this case the second trial, for the water pump experiment is

production front cover	production gasket design
low front bolt torque	use sealant
smooth finish	high back bolt torque
back-to-front torque sequence	

Complete randomization was utilized in this case since any one trial could be assembled and tested as easily as any other trial. There is no cost penalty associated with changing levels of the factors in this experiment. Due to the cost of the experiment (part cost and test stand cost), the sample size was held to one assembly per trial to initially investigate the leaking problem. The data produced by the eight unique assemblies in each trial of an L8 OA is shown in Table 4-7.

4-4-2   Die-cast piston experiment

The fourth trial data sheet would read

low copper %   high magnesium %

high zinc %   cooling water off

air cooling on

Simple repetition was utilized in this case since the alloying elements would be very difficult to change on a test-to-test basis. Once the die-cast machine was loaded with a particular alloy and temperatures stabilized, all of the pistons to be measured for hardness were collected in a short time. Since the parts were not that expensive once the machine was set up and the parts were easy to measure for hardness, three pistons were selected to represent each trial in the experiment. Each piston was measured in two locations, the dome and skirt, which then included the noise factor Z, piston position, in the experiment. The data produced by this experiment is shown in Table 4-8.

TABLE 4-7 Water Pump Experimental Results

			Factors				
	A	B	C	D	E	F	G
				Column no.			
Trial no.	1	2	3	4	5	6	7	Data leak rating
1	1	1	1	1	1	1	1	4
2	1	1	1	2	2	2	2	3
3	1	2	2	1	1	2	2	1
4	1	2	2	2	2	1	1	0
5	2	1	2	1	2	1	2	2
6	2	1	2	2	1	2	1	4
7	2	2	1	1	2	2	1	0
8	2	2	1	2	1	1	2	1

TABLE 4-8 Die-Cast Piston Experimental Results

			Factors and interactions									
	A	B	A×B D×E	C	D	E	A×E B×D					
				Column no.				Data (R_B), position (Z)				
Trial no.	1	2	3	4	5	6	7	Dome			Skirt	
1	1	1	1	1	1	1	1	71 71 72			75 74 75	
2	1	1	1	2	2	2	2	72 72 72			71 69 71	
3	1	2	2	1	1	2	2	55 55 55			68 67 68	
4	1	2	2	2	2	1	1	76 74 74			72 74 74	
5	2	1	2	1	2	1	2	78 78 77			78 78 76	
6	2	1	2	2	1	2	1	63 69 65			69 73 66	
7	2	2	1	1	2	2	1	74 70 72			72 70 74	
8	2	2	1	2	1	1	2	75 72 71			73 74 73	

4-5   Attribute versus Variable Data Considerations

Referring back to the beverage container pop-top tab failures in Sec. 2-3-3, let's assume that the actual failure rate is 0.02 percent or equal to 1 can out of every 5000 fails to open properly. The failure mode is that the tab tears away from the rivet on the top without opening the top. The customer expects the tab and rivet mechanism to be stronger than the top so that the drink may be consumed.

4-5-1   Attribute data viewpoint

To obtain the guideline of 20 occurrences of the tab failures as attribute data (it either works or it doesn't) would require a total sample size of 100,000 cans to be made and opened. The quantity of 100,000 cans times the failure rate of 0.0002 would yield 20 defective cans. This total sample size is required regardless of the number of factors that might be evaluated or which OA is used to determine how to reduce the failure rate.

4-5-2   Variable data viewpoint

Referring to Fig. 2-1, the area of overlap portrays the condition when the strength to the top exceeds the strength of the tab. Even though, on the average, the tab is stronger than the top, a certain number of tabs are weaker than tops (approximately 0.02 percent). To define these distributions, a test would have to be developed which would assess the strength of tabs and tops independently of each other. Perhaps, a test fixture could be used to evaluate the strength of tops by using a mandrel which simulated a tab to load the top until it failed. Similarly, a load cell could measure the force required to cause the tabs to fail.

Assuming for example purposes only that the variance of the strength of tops is equal to the strength of tabs, the difference between the two distributions is equal to approximately 5 standard deviations.

Two different experiments could then be conducted. One experiment would study a group of factors that would either increase the average strength of tabs or decrease the variation of strength of tabs. A second experiment would study a group of factors that would either decrease the average strength of tops or decrease the variation of strength of tops. By either increasing the difference between the two distributions or decreasing the variation of the distributions, the area of overlap would be substantially reduced or practically eliminated. Since there are physical limitations to decreasing the top strength (the can must contain the beverage pressures) and cost limitations to increasing the tab strength (thicker material, for example), the most cost-effective solution would be to reduce variation of the distributions of strength.

The two experiments could both be done with samples of no more than 32 tabs and 32 tops of various configurations. This is a tremendous reduction in the number of products to be tested compared to the attribute situation! Also, the decisions concerning what to change to improve the product would be much better from a technical aspect; the experimenter would have a greater depth of understanding of how the product actually functions.

4-6   Summary

This chapter addressed the pertinent points of collecting data from an experimental design. The main considerations of adequate sample size and randomization are used to improve the statistical validity of the experiment. Other considerations of using trial description and data sheets and the logistics of performing the tests improve the chance that proper test protocol will be followed. After tests are completed, an experimenter may observe quite easily whether or not there is improvement potential in the selected factors and levels. The contrasts of good and bad data sets are easily made when an orthogonal array is used as the experimental basis. The sample size impact of variable versus attribute data was also discussed.

The next chapter covers a major portion of the final phase of the DOE process, analysis and interpretation methods.

Problems

4-1 What purpose does an experimental trial description or data sheet serve?

4-2 What is the trial description in operational terms for trial 7 in the water pump example?

4-3 What is the trial description in operational terms for trial 3 in the die-cast piston example?

4-4 What main advantage does variable data provide compared to attribute data as experimental results?

4-5 At a 90 percent confidence level using variable data, how much change in average in terms of multiples of standard deviations will two repetitions per trial in an L8 OA detect?

4-6 At a 99 percent confidence level using variable data, how much change in average in terms of multiples of standard deviations will two repetitions per trial in an L8 OA detect?

4-7 At a 90 percent confidence level using variable data, how much change in average in terms of multiples of standard deviations will one test result per trial in an L16 OA detect?

4-8 Based on attribute data guidelines, how many samples will have to be tested when the occurrence rate is 7.5 percent (0.075)?

4-9 What advantage does multiple-class attribute data provide compared to two-class attribute data as experimental results?

4-10 What is the purpose of randomization of tests?

4-11 Under what conditions would simple repetition typically be used?

4-12 Under what conditions would complete randomization typically be used?

4-13 What do consistent results within a trial (test condition) and relatively large differences trial-to-trial indicate?

4-14 What do inconsistent results within a trial indicate?

5

Analysis and Interpretation Methods for Experiments

5-1 DOE Process Final Phase

The final phase of the DOE process is to analyze and interpret the experimental results to improve the performance characteristics of the product or process relative to customer needs and expectations. After all tests are conducted, decisions must be made concerning which parameters affect the performance of a product or process. These decisions are made with the assistance of various analytical techniques such as the

1. Observation method (Sec. 5-2)
2. Ranking methods (Sec. 5-3)
3. Column effects method (Sec. 5-4)
4. Plotting methods (Sec. 5-5)
5. Analysis of variance (Sec. 5-6)

Some of these methods for determining influential factors are subjective in nature and others are objective decision-making tools. Analysis of variance (ANOVA) will be the predominant statistical method used to interpret experimental data and make the necessary decisions since this method is the most objective. The other methods should be considered as supporting and reinforcing techniques.

It should be noted that the determination of influential factors and their relative strengths is based on the levels chosen for those factors. Regardless of the analytical method, any factor would tend to look less important if the levels chosen were closer together and any factor would tend to look more important if the levels were farther apart.

This is why the selection of levels is so critical; the interpretation depends on the test conditions.

5-2   Observation Method

This method is suggested as a preliminary interpretation method, and it is possible to use it when the response is an NB situation, but it works best with an LB or HB characteristic. The observation method is the simplest, easiest method of interpreting an experiment with an orthogonal array structure. The effort is focused on those trials having results that are very similar and having technical appeal. Usually, the portion of trials that have similar results will be 1/2 of the experiment (one strong factor), 1/4 of the experiment (two strong factors), 1/8 of the experiment (three strong factors), etc., when using two-level OAs. The results will fall into groups of 1/3 of the experiment (one strong factor), 1/9 of the experiment (two strong factors), etc., when using three-level OAs. When the most consistent, most desirable group of results has been identified, then the levels of the most important factors will be common for that group of trials. By simply observing the experimental data, determining a group of trials with technically desirable results, and finding the columns (factors and interactions) with common levels for that group of trials, the experimenter will obtain some information concerning important factors and interactions.

5-2-1   Water pump experiment

As a reminder, the seven factors mentioned in Table 2-3 were assigned to a column in an L8 OA and eight different assemblies were rated for leakage. The leak performance was monitored on a semi-variable scale with 0 representing a no-leak condition and a 5 representing the worst leak condition which had ever been observed. The experimental results are summarized in Table 4-7.

Trials 4 and 7 have the most desirable results of zero leak ratings (leak rating is an LB characteristic) from a customer's viewpoint, but what have those trials in common that might provide better pump performance? When looking at column 1, factor A, one can see that trial 4 has level A_1 and trial 7 has level A_2. Since good results were obtained in either case, it is unlikely that factor A contributes anything meaningful toward a successful result. Columns 2, 5, and 7 have common levels for trials 4 and 7 (B_2, new gasket design; E_2, smooth finish; and G_1, front-to-back torque sequence). Presumably, these factors at these levels contribute toward successful results. At this point any interaction that may exist is irrelevant (columns 2, 5, and 7 are mutually interactive); this particular combination seems to provide good results. So instead of having to control seven factors, at the most, three will have to be controlled. This would be contingent, of course, on further interpretation of this data, which may help isolate the significant factors. However, since two trials out of eight (one-fourth of the experiment) were superior in leak rating, this implies that two factors are influential in reducing water pump leaks. Which two of the three is not known based on this limited analysis?

This experiment is an excellent example for the application of the observation method of analysis.

5-2-2 Die-cast piston experiment

As a reminder, the factors for the die-cast piston experiment are shown in Table 2-3. The results are summarized in Table 4-8. The die-cast piston experiment is not as clear for interpretation by the observation method. There is only one trial, trial 5, with clearly and consistently high hardness values; this implies that there are at least three factors which influence hardness (one-eighth of the experiment is best). Two trials that are close to each are trials 4 and 5 which have factor D (level 2, water cooling off), factor E (level 1, air cooling on), and the interaction of factors D and E in common. Since there are only two factors involved in this comparison, these are probably the strongest two factors.

5-3 Ranking Method

An extension of the observation method is to rank all of the results, from the best to the worst, along with the corresponding trial conditions. With this method, consistency of the levels at the ends of the goodness to badness scale are of interest. If there is a very strong relationship of the factor to the quality characteristic of interest, then all of the first levels will fall on the goodness end and all of the second levels on the badness end of the scale (or vice versa). There is a statistical confidence based on this condition which can be determined by nonparametric statistical methods. This method of determining confidence is beyond the scope of this text.

A limitation of this method is that interaction columns cannot be evaluated by the ranking method (see the die-cast piston example).

5-3-1 Water pump experiment

The water pump experimental results are ranked from best to worst along with the corresponding trial levels, as shown in Table 5-1. In this experiment, there is a very strong relationship of factor B, gasket design, with the leak rating. All of the factor B_2 levels fall at the good end of the leakage scale and all of the factor B_1 levels fall at the bad

TABLE 5-1   Water Pump Experiment (Ranked Results)

			Factors					
	A	B	C	D	E	F	G	
				Column no.				
Trial no.	1	2	3	4	5	6	7	Data leak rating
4	1	2	2	2	2	1	1	0
7	2	2	1	1	2	2	1	0
3	1	2	2	1	1	2	2	1
8	2	2	1	2	1	1	2	1
5	2	1	2	1	2	1	2	2
2	1	1	1	2	2	2	2	3
1	1	1	1	1	1	1	1	4
6	2	1	2	2	1	2	1	4
		1st			2nd			

end of the scale. This is the only factor with this strong a relationship. Factor E has moderate strength since two 2s are at the good end and two 1s at the bad end of the scale. Note, also, that the relationship for factor E remains intact, within each of the factor B levels indicating secondary strength to factor B. The new gasket design in conjunction with a smooth pump housing finish should provide a leakproof unit.

5-3-2   Die-cast piston experiment

The die-cast piston experiment ranked results are shown in Table 5-2. The results are ranked in order of the average of the results within a

TABLE 5-2   Die-Cast Piston Experiment (Ranked Results)

			Factors and interactions						
	A	B	A×B D×E	C	D	E	A×E B×D		
				Column no.				Data (R_B), position (Z)	
Trial no.	1	2	3	4	5	6	7	Dome	Skirt
5	2	1	2	1	2	1	2	78 78 77	78 78 76
4	1	2	2	2	2	1	1	76 74 74	72 74 74
1	1	1	1	1	1	1	1	71 71 72	75 74 75
8	2	2	1	2	1	1	2	75 72 71	73 74 73
7	2	2	1	1	2	2	1	74 70 72	72 70 74
2	1	1	1	2	2	2	2	72 72 72	71 69 71
6	2	1	2	2	1	2	1	63 69 65	69 73 66
3	1	2	2	1	1	2	2	55 55 55	68 67 68
					2nd	1st			

trial; trial 5 has the highest average for the six measurements and trial 3 has the lowest average. Factor E, air cooling, has a strong relationship with piston hardness. All of the factor E_1 levels fall at the high end of the hardness scale and all of the factor E_2 levels fall at the low end of the scale. Factor D, water cooling, has moderate strength since two 2s are at the high end of the hardness scale and two 1s are at the low end of the scale.

Interactions generally cannot be evaluated with this technique. For example, the $D \times E$ interaction has two 2s at the high end of the scale, and two 2s at the low end of the scale, which is difficult to interpret. The column effects or ANOVA methods will show the relative strengths of the interaction columns.

5-3-3 Commentary on observation and ranking methods

Both the observation and ranking methods have an interesting property when using two-level OAs as the experimental basis, that is, to divide the results into groups which are consistent within the group. With two-level arrays, if one factor is really the important factor, then there will tend to be two groups of data: one group associated with the low level and a second group associated with the high level. If two factors are really important, then there will tend to be four groups: two trials in an L8 may be the best results or four trials in an L16 may be the best results. Because of a two-level design, the groups should be made up of an even number of tests.

If some odd number of trials appears in a group for two-level arrays (or an even number of trials in a group for three-level arrays), some note should be made of this. This may be an indication of either the measurement system contributing a sizable portion of the total variation or other strong factors that are not included in the experiment changing the results in a random fashion.

5-4 Column Effects Method

This approach is used by Taguchi as a simplified ANOVA to subjectively point out columns which have large influences on the response. The sum of the data associated with the first level is subtracted from the sum of the data associated with the second level for each column in an OA.

Three pieces of information are generated from the column effects analysis:

1. Which factors make a difference
2. The relative importance of those factors

3. Which direction for levels of those factors will lead to further improvement

The magnitudes of the differences are compared to each other to find the relatively large effects. The relative magnitudes (the plus or minus sign shows positive or negative correlation with level numbers, respectively) indicate the relative power of the factors in affecting the results. The strongest factors or interactions will have the largest differences. The level sum that is more desirable from a technical viewpoint indicates whether lower levels or higher levels of the influential factors potentially would have even better results (this will have to be verified by tests, of course).

5-4-1 Water pump experiment

Table 5-3 shows a column effects analysis method summary. Looking at the difference row, factor B has the largest effect (-11), factor E the second largest effect (-5), and all other factors have a weak effect or no measurable effect.

The observation method indicated three possible factors as important $(B, E,$ and $G)$, while the ranking and column effects methods indicated only two factors as important $(B$ and $E)$. The observation method used a small portion of the complete data set, while the ranking and column effects methods used all the data. One factor may drop out of overall contention as an influential factor when all the data is considered. These methods do, however, all indicate the same

TABLE 5-3   Water Pump Experiment (C-E Method)

	A	B	C	D	E	F	G	
				Factors				
				Column no.				
Trial no.	1	2	3	4	5	6	7	Data leak rating
1	1	1	1	1	1	1	1	4
2	1	1	1	2	2	2	2	3
3	1	2	2	1	1	2	2	1
4	1	2	2	2	2	1	1	0
5	2	1	2	1	2	1	2	2
6	2	1	2	2	1	2	1	4
7	2	2	1	1	2	2	1	0
8	2	2	1	2	1	1	2	1
Sum$_1$	8	13	8	7	10	7	8	
Sum$_2$	7	2	7	8	5	8	7	
Difference	-1	-11	-1	1	-5	1	-1	

TABLE 5-4   Die-Cast Piston Experiment (C-E Method)

			Factors and interactions										
			$A \times B$				$B \times D$						
	A	B	$D \times E$	C	D	E	$B \times D$						
				Column no.				Data (R_B), position (Z)					
Trial no.	1	2	3	4	5	6	7	Dome			Skirt		
1	1	1	1	1	1	1	1	71	71	72	75	74	75
2	1	1	1	2	2	2	2	72	72	72	71	69	71
3	1	2	2	1	1	2	2	55	55	55	68	67	68
4	1	2	2	2	2	1	1	76	74	74	72	74	74
5	2	1	2	1	2	1	2	78	78	77	78	78	76
6	2	1	2	2	1	2	1	63	69	65	69	73	66
7	2	2	1	1	2	2	1	74	70	72	72	70	74
8	2	2	1	2	1	1	2	75	72	71	73	74	73
Sum$_1$	1677	1735	1735	1703	1649	1785	1719						
Sum$_2$	1740	1682	1682	1714	1768	1632	1698						
Difference	63	−53	−53	11	119	−153	−21						

relatively strong factors and relationships for improving the sealability of the engine water pump. No statistical decision criteria were used for any of these cursory analysis techniques. They should be used in conjunction with ANOVA for comprehensive evaluation.

5-4-3   Die-cast piston experiment

The column effects summary is shown in Table 5-4. Overall, two factors stand out as strong: factor E with a difference of −153 and factor D with a difference of +119. Factors A, B, and the $A \times B / D \times E$ interaction are questionable as to their effect on piston hardness. More sophisticated methods, such as ANOVA, will be required to determine their statistical significance.

In this case, there is total agreement of the observation, ranking, and column effects method for the two strongest factors.

5-5   Plotting Methods

The plotting of experimental results should be done in as many ways as are meaningful:

1. By levels of influential factors

2. By levels of combinations of influential factors to assess interaction possibilities

3. By order of actual test

5-5-1   Plotting by levels

The column effects, and later ANOVA, indicate what to plot. To plot the effect of influential factors, the average result for each level must be calculated first. The sum of the data associated with each level in the OA column divided by the number of tests (data points) for that level will provide the appropriate averages. This applies to factors with two or more levels.

Plots may be made with equal increments between levels on the horizontal axes of the graphs to show the relative strengths of the factors. The factor's strength is directly proportional to the slope of the graph. An actual scale may also be used for the horizontal axis of continuous factors to graphically interpolate or extrapolate for predictions of other levels. With the actual scale, however, the relative comparison of slopes is not meaningful.

Water pump experiment. The two influential factors are gasket design, factor B, and pump finish, factor E. Using the level 1 and level 2 sums in the column effects table, the average leak rating for those factors may be determined. The average leak rating for the B_1 and B_2 levels is 3.25 and 0.5, respectively. The average leak rating for the E_1 and E_2 levels is 2.5 and 1.25, respectively. The plots of these factors are shown in Fig. 5-1.

Die-cast piston experiment. The two influential factors are air cooling, factor E, and water cooling, factor D. Using the level 1 and level 2 sums in the column effects table, the average hardness for those factors may be determined. The average hardness for the E_1 and E_2 lev-

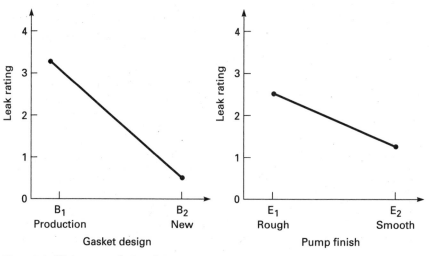

Figure 5-1 Water pump factor plots.

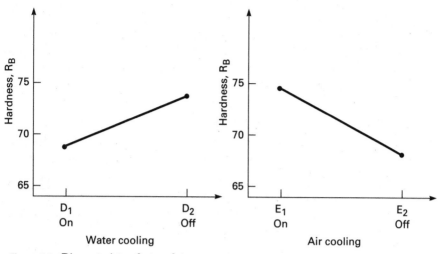

Figure 5-2 Die-cast piston factor plots.

els is 74.38 and 68.00, respectively. The average hardness for the D_1 and D_2 levels is 68.71 and 73.67, respectively. The plots of these factors are shown in Fig. 5-2.

5-5-2 Plotting by combinations of levels (interactions)

This procedure is somewhat more complex, but can be accomplished by referring back to the OA and the raw data associated with each trial. In an L8 OA example using factor A assigned to column 1 and factor B assigned to column 2, the $A \times B$ interaction plot may be made in the following manner. The average of the results for each of the factor and level combinations must be calculated. In this example, there are four combinations with two trials representing each of those combinations. Trials 1 and 2 represent the A_1B_1 combination, for example. All of the data associated with trials 1 and 2 must be summed and divided by the number of tests (data points) that made up that sum to obtain the desired average. Each of the other combinations may be calculated in a similar manner using the appropriate trials. A standard, unmodified, two-level OA will automatically have an equal number of trials representing the four combinations regardless of which columns are involved in the interaction of interest.

The interpretation is determined by the parallelism of the plotted lines. If the lines are parallel or almost parallel, this indicates there is no meaningful interaction taking place between the plotted factors. No interaction means the effect of the first factor is the same regardless of the level of the second factor and vice versa. The greater the skew between the lines, the greater the strength of the interaction between the factors. An interaction means the effect of the first factor depends on which level of the second factor is used and vice versa.

Water pump experiment. The averages of the four combinations of factors B and E are

$$B_1E_1 = \frac{(4 + 4)}{2} = 4$$

$$B_1E_2 = \frac{(3 + 2)}{2} = 2.5$$

$$B_2E_1 = \frac{(1 + 1)}{2} = 1$$

$$B_2E_2 = \frac{(0 + 0)}{2} = 0$$

The plot showing the interaction of factors B and E is shown in Fig. 5-3. The lines representing the rough and smooth finishes are almost parallel which indicates that there is little if any meaningful interaction between factors B and E. No interaction means that the factor B effect

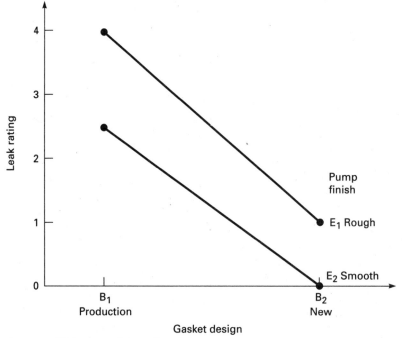

Figure 5-3 Water pump interaction plot.

is the same regardless of the level of factor E and vice versa. Factor B causes approximately a 2.75 drop in leak rating at both levels of factor E and factor E causes approximately a 1.25 drop in leak rating at both levels of factor B.

Die-cast piston experiment.   The plot of factors D and E is shown in Fig. 5-4. There is a slight angle between the two lines, which indicates a slight interaction is present between factors D and E.

5-5-3   Plotting by order of test

Plotting in order of actual test will reveal any strong trends, up or down, that may have existed during the execution of the experiment. The trends would have to be caused by some unknown, uncontrolled factor(s) that varied during the experiment. Hopefully, the data will show no correlation (no trend up or down) with the test order, which would disturb the interpretation of results. If a trend up or down is detected, the data should be corrected to account for the trend and analyzed after correction.

To correct the data for a trend so that it appears as a horizontal pattern is very simple. A trend line is drawn through the data with respect to actual test order. The data values should have the difference between the trend line and $\overline{T}$ subtracted from the data in the area where the trend line is above $\overline{T}$. The difference between the

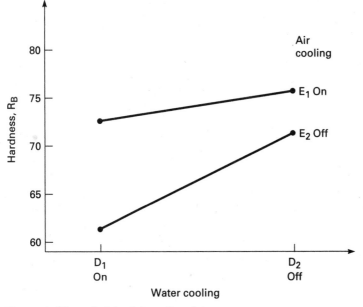

Figure 5-4   Die-cast piston interaction plot.

trend line and $\overline{T}$ should be added to the data in the area where the trend line is below $\overline{T}$. This removes the trend effect from the data, which reduces error variation and makes ANOVA more sensitive to differences in averages that may exist.

If a trend should exist, it is recommended that the experimenter pursue the factor(s) that may have produced the trend.

5-6   Analysis of Variance (ANOVA)

This method was developed by Sir Ronald Fisher in the 1930s as a way to interpret the results from agricultural experiments. ANOVA is not a complicated method and has a lot of mathematical beauty associated with it. ANOVA is a statistically based, objective decision-making tool for detecting any differences in average performance of groups of items tested. The decision, rather than using pure judgment, takes variation into account.

The discussion of ANOVA will start with a very simple case, no-way ANOVA, and build up to more comprehensive situations, three-way ANOVA, in this chapter. Also, ANOVA will be applied to experimental situations utilizing orthogonal arrays, although this analysis method can be used with any set of data that has some structure. The experimental designs and subsequent analyses are intrinsically tied to one another.

5-7   No-Way ANOVA

No traditional statistics book will mention no-way ANOVA, but it is the simplest situation to analyze and has some practical value which will be discussed in later chapters.

5-7-1   Sums of squares

No-way ANOVA begins with a set of experimental data, real or, as in this case, assumed for simplification. Imagine an engineer is sent to a production line to sample a set of windshield washer pumps for the purpose of measuring flow rate. The data collected could be like that in Table 5-5.

Although this data is entirely fictitious and put into single digits to

TABLE 5-5   Low-Capacity Pump Flow Rate

Pump no.	1	2	3	4	5	6	7	8
Flow rate (oz/min)*	5	6	8	2	5	4	4	6

*1 oz/min = 0.473 ml/s.

simplify calculations, these observations are quite possible. ANOVA would mathematically be carried out in an identical manner regardless of the data actually collected or the units of the data points. Many of the examples in this book are simplified in this manner to demonstrate the analysis method rather than to test the mathematical ability of the reader.

Analysis of variance is a mathematical technique which breaks total variation down into accountable sources; total variation is decomposed into its appropriate components. No-way ANOVA, the simplest case, breaks total variation down into only two components:

1. The variation of the average (or mean) of all the data points relative to zero

2. The variation of the individual data points around the average (traditionally called experimental error)

Some notation is necessary to demonstrate the calculation method:

y = observation, response, data
y_i = ith response; y_3 = 8 oz/min
N = total number of observations
T = sum of all observations
$\overline{T}$ = average of all observations = $T/N = \overline{y}$

In this case

$$N = 8 \qquad T = 40 \text{ oz/min} \qquad \overline{T} = 5.0 \text{ oz/min}$$

The word "observation" has a particular connotation. The true flow rate of any given pump is actually unknown; it is only estimated through the use of some flow meter. There will be some unknown measurement error present, but a flow rate will nonetheless be observed and accepted as the pump's performance under the conditions of the test. Also, these pumps were randomly selected from a production line which ostensibly manufactures identical pumps; however, there will be slight differences from pump to pump, causing a portion of the pump-to-pump variation in performance. Why are not all of the pump flow rates identical? Would we expect them to be?

No-way ANOVA can be illustrated graphically, which enhances the understanding. A plot of the data appears in Fig. 5-5. The magnitude of each observation can be represented by a line segment extending from zero to the observation. These line segments can be divided into two portions: one portion attributed to the mean and one portion attributed to error. Error includes the measurement errors plus the effects of pump characteristics that are varying slightly from pump to

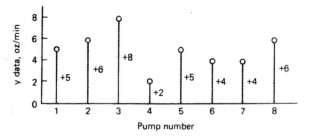

Figure 5-5 Low-capacity pump performance; total variation.

pump.

The magnitude of the line segment due to the mean is indicated by extending a line from the average value to zero, as shown in Fig. 5-6. The magnitude of the line segment due to error is indicated by the difference of the average value from each observation, as shown in Fig. 5-7.

Note that the summation of the line segments due to the mean equals 40.0, which is also equal to the sum of the raw data points. The summation of the error term around the average is equal to zero. This in itself is not very informative; however, there is a mathematical operation to be performed which allows a clearer picture to devel-

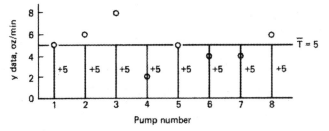

Figure 5-6 Low-capacity pump performance; variation due to mean.

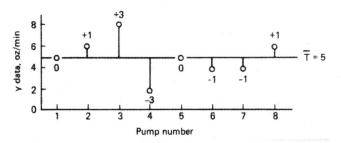

Figure 5-7 Low-capacity pump performance; variation due to error.

op. The magnitudes of each of the line segments can be squared and then summed to provide a measure of the total variation present. Referring to Fig. 5-5, the total sums of squares (total variation) is then computed:

$$SS_T = \text{total sums of squares}$$
$$SS_T = 5^2 + 6^2 + 2^2 + 5^2 + 4^2 + 4^2 + 6^2$$
$$SS_T = 222.0$$

The magnitude of the portion of the line segment due to the mean can also be squared and summed. Referring to Fig. 5-6, the variation due to the mean is then computed:

$$SS_m = \text{sums of squares due to the mean}$$
$$SS_m = N(\overline{T})^2$$

But $\overline{T} = T/N$, so

$$SS_m = N\left[\frac{T}{N}\right]^2 = \frac{T^2}{N}$$

$$SS_m = \frac{40^2}{8} = 200.0$$

The portion of the magnitude of the line segment due to error can be squared and summed to provide a measure of the variation around the average value. Referring to Fig. 5-7, the variation due to error is

$$SS_e = \text{error sums of squares}$$
$$SS_e = 0^2 + 1^2 + 3^2 + (-3)^2 + 0^2 + (-1)^2 + (-1)^2 + 1^2$$
$$SS_e = 22.0$$

Note that

$$222.0 = 200.0 + 22.0$$

This demonstrates (not a proof) a basic property of ANOVA. The total sums of squares is equal to the sum of the sums of squares due to the known components. In this case,

$$SS_T = SS_m + SS_e$$

In this method, the total variation can be decomposed into two sources with the appropriate share apportioned to each source. The formulas for the sums of squares can be written generally

$$SS_T = \sum_{i=1}^{N} y_i^2$$

which is the summation of the squares of each observation from $i = 1$ to N, and

$$SS_m = \frac{T^2}{N}$$

which is equivalent to the summation of the square of the portion of each observation due to the mean for $i = 1$ to N.

$$SS_e = \sum_{i=1}^{N} (y_i - \overline{T})^2$$

which is the summation of the square of the differences of each observation from the mean.

Here, the error component was actually calculated, but this was not really necessary. The ANOVA method used states that

$$SS_T = SS_m + SS_e$$

so

$$SS_e = SS_T - SS_m = 222 - 200 = 22$$

Practice problem 1. Another model of windshield washer pump is sampled from production (see Table 5-6). It is recommended that the reader perform a no-way ANOVA on this set of observations. Answers are in App. A.

5-7-2   Degrees of freedom

To complete the ANOVA calculations, one other element must be considered: degrees of freedom. A degree of freedom in a statistical sense is associated with each piece of information that is estimated from the data. For instance, the mean (average) is estimated from all the data and requires one degree of freedom (d.f.) for that purpose. Another way to think of the concept of degrees of freedom is to allow 1 d.f. for each fair (independent) comparison that can be made in the data. Only one fair comparison can be made between the mean of all the data (there is only one mean) and zero, the reference point; 5.0 oz/min has meaning only when thought of in terms of the reference point 0.0

TABLE 5-6   High-Capacity Pump Flow Rate

Pump no.	1	2	3	4	5	6	7	8
Flow rate (oz/min)	15	16	18	12	15	14	14	16

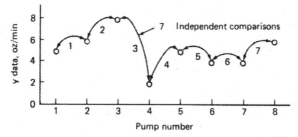

Figure 5-8 Low-capacity pump performance; degrees of freedom.

oz/min. Therefore, there is only 1 d.f. associated with the mean, which is only one piece of information.

This concept of independent comparisons also applies to the degrees of freedom associated with the error estimate. With reference to the eight observations, there are seven independent comparisons that can be made to estimate the variation in the data. Data point 1 can be compared to data point 2, 2 to 3, 3 to 4, etc., as shown in Fig. 5-8. In this data set there exist only seven independent comparisons. Data point 1 compared to data point 3 is not another independent comparison; that comparison is dependent on the comparison of data point 1 to 2 and 2 to 3.

Similar to sums of squares, a summation can be made for degrees of freedom, where

$$SS_T = SS_m + SS_e$$

let v = degrees of freedom
v_T = total degrees of freedom
v_m = degree of freedom associated with the mean (always 1)
v_e = degrees of freedom associated with error

$$v_T = v_m + v_e$$
$$8 = 1 + 7$$

The total degrees of freedom equals the total number of observations in the data set for this method of ANOVA.

TABLE 5-7   No-Way ANOVA Summary

Source	SS	d.f.
Mean	200	1
Error	22	7
Total	222	8

An ANOVA table can now be constructed which summarizes all the information for the low-capacity pump data set (see Table 5-7).

5-7-3   Variance due to error

One other descriptive statistic that can be calculated from the ANOVA table is the variance V. Error variance, usually termed just variance, is equal to the sums of squares for error divided by the degrees of freedom for error; in the example,

$$V_e = \frac{SS_e}{v_e}$$

$$V_e = \frac{22}{7} = 3.14$$

By definition, the standard deviation is equal to the square root of variance.

$$S = \sqrt{V}$$

S = sample standard deviation
σ = population standard deviation (sigma)

S is an estimate of σ, the true but unknown standard deviation.

This calculation is identical to the standard deviation calculation done by an electronic calculator. The formula used is

$$S = \sqrt{\frac{\sum_{i=1}^{N}(y_i - \bar{y})^2}{N-1}} \qquad \text{standard deviation}$$

$$S^2 = V = \frac{\sum_{i=1}^{N}(y_i - \bar{y})^2}{N-1} \qquad \text{variance}$$

Notice that the numerator is the same calculation as for the sums of squares due to error, and that the denominator is the degrees of freedom associated with error variation. Although the preceding formula is much faster than ANOVA for calculating error variance in the example given, when the experimental situations become more complex, ANOVA will become the faster method.

Error variance is a measure of the variation due to all the uncontrolled parameters, including measurement error involved in a particular experiment (set of data collected).

5-7-4 Method of least squares

This ANOVA method is based on a least squares approach; the error variance is equal to the minimum value of the sums of squares about some reference value divided by the degrees of freedom for error. Notice that the practice problem had exactly the same error variance as the example problem, even though the zero reference line was 10 units further away (a value of 10 was added to all the original observations). It so happens that the error sums of squares are minimized when the reference line is the average value of the observations.

5-8 One-Way ANOVA

One-way ANOVA is the next most complex ANOVA to conduct. This situation considers the effect of one controlled parameter upon the performance of a product or process, in contrast to no-way ANOVA, where no parameters were controlled.

5-8-1 Product development situation

Again, an imaginary, yet potentially very real, situation is presented as a typical case when one-way ANOVA is applicable. Imagine the same engineer as before is charged with the task of establishing the fluid velocity generated by the windshield washer pumps. Obviously, if the fluid velocity is too low, the fluid will merely dribble out, and if too high, air movement past the windshield will not be able to distribute the cleaning fluid adequately to satisfy the driver of the car. The engineer proposes a test of three different orifice areas to determine which may give a proper fluid velocity.

Before the test data is collected some notation is in order to simplify the mathematical discussion.

A = factor under investigation (outlet orifice area)
A_1 = 1st level of orifice area = 0.0015 in^2 (0.97 mm^2)
A_2 = 2d level of orifice area = 0.0030 in^2 (1.94 mm^2)
A_3 = 3d level of orifice area = 0.0045 in^2 (2.90 mm^2)

The same symbol for the level designation will be used to denote the sum of responses for that test condition.

$\underline{A_i}$ = sum of observations under A_i level
$\overline{A_i}$ = average of observations under A_i level = A_i / n_{A_i}
$\underline{T}$ = sum of all observations
$\overline{T}$ = average of all observations = T/N
n_{A_i} = number of observations under A_i level
N = total number of observations

TABLE 5-8   Pump Velocity Data (First Experiment)

Level	Area, in^2	Velocity, ft/s				Total
A_1	0.0015	2.2	1.9	2.7	2.0	8.8
A_2	0.0030	1.5	1.9	1.7	*	5.1
A_3	0.0045	0.6	0.7	1.1	0.8	3.2
		Grand Total				17.1

*Dropped pump and destroyed, no data.
NOTE: 1 ft/s = 0.30 m/s.

k_A = number of levels of factor A

With this notation in mind, the engineer constructs four pumps with a given orifice area, making a total of 12 to test. The test data is shown in Table 5-8. Using the notation system, A_1 represents 0.0015 in^2 orifice area and also the sum of observations of the pump with 0.0015 in^2 orifice area, which is 8.8 ft/s. Then,

$$A_1 = 8.8 \text{ ft/s} \quad n_{A_1} = 4 \quad \overline{A}_1 = 2.2 \text{ ft/s}$$

$$A_2 = 5.1 \text{ ft/s} \quad n_{A_2} = 3 \quad \overline{A}_2 = 1.7 \text{ ft/s}$$

$$A_3 = 3.2 \text{ ft/s} \quad n_{A_3} = 4 \quad \overline{A}_3 = 0.8 \text{ ft/s}$$

$$T = 17.1 \text{ ft/s} \quad N = 11 \quad \overline{T} = 1.6 \text{ ft/s}$$

$$k_A = 3$$

5-8-2   Sums of squares (one-way ANOVA)

Two methods can be used in ANOVA to complete the calculations:

1. Including the mean (Sec. 5-8-3)

2. Excluding the mean (Sec. 5-8-6)

5-8-3   Method 1 (including the mean)

As before, the total variation can be decomposed into its appropriate components.

1. The variation of the average (mean) of all observations relative to zero

2. The variation of the average (mean) of observations under each factor level around the average of all observations

3. The variation of the individual observations around the average of observations under each factor level

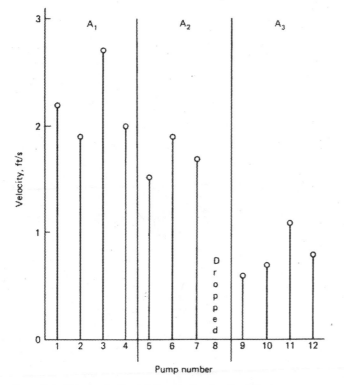

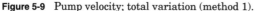

Figure 5-9 Pump velocity; total variation (method 1).

The calculations are identical to the no-way ANOVA example, with the exception of the component of variation due to factor A, outlet orifice area. Graphically, this can be seen quite clearly in Fig. 5-9.

$$SS_T = \sum_{i=1}^{N} y_i^2$$

$$SS_T = 2.2^2 + 1.9^2 + 2.7^2 + \dots + 0.8^2 = 31.190$$

Figure 5-10 shows graphically the line segment which is proportional to the variation due to the mean.

$$SS_m = N(\overline{T})^2 = \frac{T^2}{N}$$

$$SS_m = \frac{17.1^2}{11} = 26.583$$

This calculation is based on the format used in previous sums of

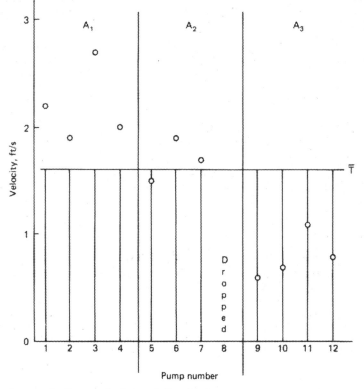

Figure 5-10 Pump velocity; variation due to mean.

squares calculations. The magnitude of the line segments due to each level of factor A is squared and summed. For instance, the length of line segment due to level A_1 is $(\overline{A}_1 - \overline{T})$. There are four observations under level A_1 condition. The same type of information is used for the other levels of factor A. Figure 5-11 shows this graphically.

$$SS_A = n_{A_1}(\overline{A}_1 - \overline{T})^2 + n_{A_2}(\overline{A}_2 - \overline{T})^2 + n_{A_3}(\overline{A}_3 - \overline{T})^2$$
$$SS_A = 4(0.64545)^2 + 3(0.14545)^2 + 4(-0.75454)^2$$
$$SS_A = 4.007$$

This calculation method is very tedious, but it is mathematically equivalent to

$$SS_A = \left[\sum_{i=1}^{k_A}\left(\frac{A_i^2}{n_{A_i}}\right)\right] - \frac{T^2}{N}$$

(See App. E, proof 2.)

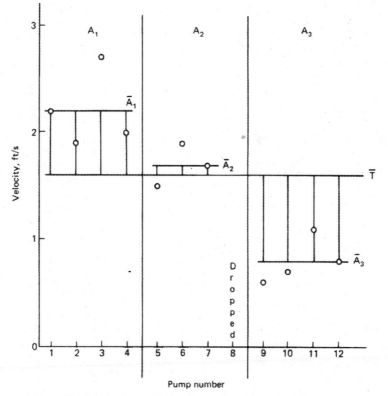

Figure 5-11   Pump velocity; variation due to factor A.

$$SS_A = \frac{8.8^2}{4} + \frac{5.1^2}{3} + \frac{3.2^2}{4} - \frac{17.1^2}{11} = 4.007$$

Figure 5-12 shows the remaining portion of the line segment, which is proportional to the variation due to error.

$$SS_e = \sum_{j=1}^{k_A} \sum_{i=1}^{n_{A_j}} (y_{ij} - \overline{A}_j)^2$$

$$SS_e = 0^2 + (-0.3)^2 + 0.5^2 + (-0.2)^2 + 0.2^2 + 0^2 + (-0.2)^2 + (-0.1)^2 + 0.3^2 + 0^2 = 0.600$$

Error variation is again based on the method of least squares, but in one-way ANOVA the least squares are evaluated around the average for each level of the controlled factor. Error variation is the uncontrolled variation within the controlled groups. Again, the total variation can be accounted for mathematically.

$$SS_T = SS_m + SS_A + SS_e$$

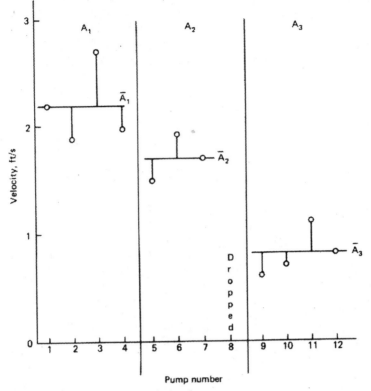

Figure 5-12 Pump velocity; variation due to error.

$$31.190 = 26.583 + 4.007 + 0.600$$

5-8-4 Degrees of freedom
(including the mean)

In method 1, including the mean, the total degrees of freedom is the sum of degrees of freedom for the mean, for factor A, and for error.

$$v_T = v_m + v_A + v_e$$
$$v_T = N = 11$$
$$v_A = k_A - 1 = 3 - 1 = 2$$
$$v_e = v_T - v_m - v_A = 11 - 1 - 2 = 8$$

The degrees of freedom for a controlled factor such as A follow the same concept of 1 d.f. for each fair comparison. $\overline{A}_1$ compared to $\overline{A}_2$ has 1 d.f. and $\overline{A}_2$ compared to $\overline{A}_3$ has 1 d.f. Also apparent in the data set

TABLE 5-9   Factor and Error Degrees of Freedom

Level	Velocity, ft/s	$\overline{A}$		
A_1	2.2 1.9 2.7 2.0 1 2 3	2.2	1	
A_2	1.5 1.9 1.7 * 4 5	1.7		2 d.f. for factor A
A_3	.6 .7 1.1 .8 6 7 8	.8	2	

8 d.f. for error

*Dropped pump and destroyed it, no data.

are the eight degrees of freedom for error, eight fair comparisons within groups. Error estimation is always done within groups. For instance, the comparison of 1.5 to 0.6 is one estimate of the A_2 to A_3 effect, not an estimate of error. Table 5-9 shows the factor and error degrees of freedom.

5-8-5   ANOVA summary table (including the mean)

The ANOVA summary is shown in Table 5-10. Now, as well as error variance being estimated, the variance due to factor A is estimated. The meaning of this estimate and the relevance will be discussed later.

5-8-6   Method 2 (excluding the mean)

It has been shown in a previous example that the variation due to the mean does not affect the calculations for the variation due to error, and for edification, neither does it affect the calculations for the factor effects. In most experimental situations, with the exception of lower-is-better characteristics, the variation due to the mean has no practical value. In a lower-is-better situation, the variation due to the mean is a measure of how far the average is from zero and how successful

TABLE 5-10   One-Way ANOVA Summary (Method 1)

Source	SS	v	V
m	26.583	1	26.583
A	4.007	2	2.004
e	0.600	8	0.075
T	31.190	11	

the factors might be in reducing the average to zero.

In the situation of the exclusion of the mean from the ANOVA calculations, total variation may be decomposed into

1. The variation of the average of observations under each factor level around the average of all observations
2. The variation of the individual observations around the average of observations under each factor level

Again, graphically, this can be demonstrated as shown in Fig. 5-13. The same concept of summing the squares of the magnitudes of the various line segments is applied in method 2 also.

$$SS_T = \sum_{i=1}^{N} (y_i - \overline{T})^2$$

$$SS_T = 0.645^2 + 0.345^2 + 1.145^2 + \ldots + (-0.755)^2$$

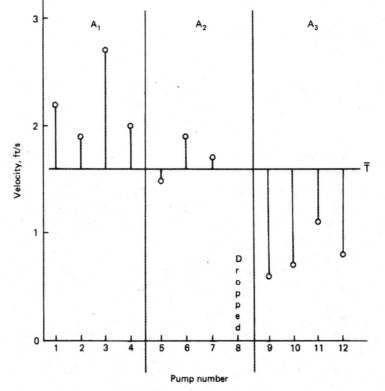

Figure 5-13 Pump velocity; total variation (method 2).

$$SS_T = 4.607$$

Mathematically, this is equivalent to

$$SS_T = \left[\sum_{i=1}^{N} y_i^2 \right] - \frac{T^2}{N} \tag{5-1}$$

(See App. E, proof 3). From previous calculations,

$$SS_T = 31.190 - 26.583 = 4.607$$

Equation 5-1 will be used primarily as the definition for total variation.

The variation due to factors is calculated identically to method 1; refer to Fig. 5-11.

$$SS_A = \left[\sum_{i=1}^{k_A} \left(\frac{A_i^2}{n_{A_i}} \right) \right] - \frac{T^2}{N}$$

The variation due to error is also identical to method 1; refer to Fig. 5-12.

$$SS_e = \sum_{j=1}^{k_A} \sum_{i=1}^{n_{A_j}} (y_i - \overline{A}_j)^2$$

5-8-7   Degrees of freedom (excluding the mean)

In method 1, the degrees of freedom formula was written

$$v_T = v_m + v_A + v_e$$

$$v_m = 1 \text{ (always)}; \qquad v_T = N$$

In method 2, the mean being excluded, the degree of freedom for the mean is subtracted from both sides of the equation:

$$N = 1 + v_A + v_e$$

$$N - 1 = v_A + v_e$$

A new definition used in method 2 is

$$v_T = N - 1$$

Therefore,

$$v_T = v_A + v_e$$

$$v_T = N - 1 = 11 - 1 = 10 \qquad \text{for the total d.f.}$$

$$v_A = k_A - 1 = 3 - 1 = 2 \qquad \text{for the d.f. for factor } A$$

$$v_e = v_T - v_A = 10-2 = 8 \qquad \text{for the d.f. for error}$$

5-8-8 ANOVA summary table (excluding the mean)

The ANOVA summary table for method 2 appears in Table 5-11. The identical variance values are calculated for factor A and error in one-way ANOVA using method 1 or method 2. The value for the mean is disregarded in method 2, which is the most popular method. Only in a case where the performance parameter is a lower-is-better characteristic would the variance due to the mean be relevant; this provides a measure of how effective some factor might be in reducing the average to zero.

Practice problem 2. Imagine a similar experiment on the higher-capacity pump using generally larger orifices, with the results shown in Table 5-12. It is again recommended that the reader calculate the items necessary for the ANOVA table using method 2. The results are shown in Table 5-13.

5-8-9 Central limit theorem

TABLE 5-11   One-Way ANOVA Summary (Method 2)

Source	SS	v	V
A	4.007	2	2.004
e	0.600	8	0.075
T	4.607	10	

TABLE 5-12   Pump Velocity Data (Second Experiment)

Level	Area, in²	Velocity, ft/s				Total
A_1	0.008	0.8	1.2	0.9	1.3	4.2
A_2	0.006	1.5	1.4	1.2	1.8	5.9
A_3	0.004	2.3	1.9	1.9	2.1	8.2
		Grand Total				18.3

TABLE 5-13   One-Way ANOVA Summary

Source	SS	v	V
A	2.0150	2	1.008
e	0.4675	9	0.052
T	2.4825	11	

The variance of factor A, shown in the ANOVA tables, is an estimate of error variance based upon the variation of the averages $\overline{A}_1, \overline{A}_2$, and $\overline{A}_3$ around the grand average $\overline{T}$. The error variance shown in the ANOVA tables is based upon the variation of the actual individual observations within a controlled group (the data points under condition A_1, for example). This situation is based on one of the fundamental theorems of statistics, the central limit theorem (CLT). The CLT has three tenets:

1. Sample averages tend to be normally distributed regardless of the distribution of the individuals.
2. The average of the distribution of sample averages will approach the average of the distribution of the individuals.
3. The variance of the sample averages is less than the variance of the distribution of the individuals.

All of the preceding are predicated on the fact that one population is being sampled; therefore, it has a constant average and constant standard deviation.

The formula that describes the third and most important tenet is

$$\sigma_{\bar{y}}^2 = \frac{\sigma_y^2}{n} \qquad (5\text{-}2)$$

This formula states that the variance of sample averages will be equal to the variance of the individuals divided by the sample size used to obtain the sample averages. Graphically, this appears in Fig. 5-14. Formula (5-2) may be rewritten

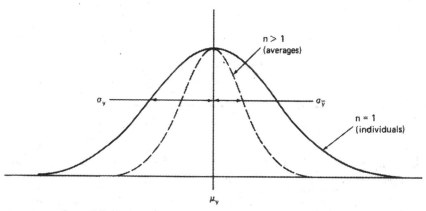

Figure 5-14 Central limit theorem.

$$\sigma_y^2 = n\sigma_{\bar{y}}^2$$

As estimates of σ, use S

$$S_y^2 = n(S_{\bar{y}})^2$$

5-8-10   Variance calculation using CLT

This formula can be used to make an estimate of individual variance by taking the variance of the sample averages and multiplying by the sample size (equal sample size for all samples is required). This property of the CLT can be applied to the practice problem, where all levels had a sample size of $n = 4$.

The variance of sample averages $\bar{A}_1$, $\bar{A}_2$, and $\bar{A}_3$ can be calculated using the variance formula modified for averages. Rather than this formula for individuals,

$$S_y^2 = \frac{\sum_{i=1}^{N} (y_i - \bar{T})^2}{N - 1}$$

substitute k_A for N and $\bar{A}$ for y.

$$S_A^2 = \frac{\sum_{i=1}^{k_A} (\bar{A}_i - \bar{T})^2}{k_A - 1}$$

$$S_A^2 = \frac{(1.050 - 1.525)^2 + (1.465 - 1.525)^2 + (2.050 - 1.525)^2}{3 - 1}$$

$$S_A^2 = 0.252$$

But

$$S_y^2 = n(S_A)^2 \qquad \text{where } n = 4$$
$$S_{y_1}^2 = 4(0.252) = 1.008$$

Notice that this value is equal to the variance of factor A in the ANOVA table, but it is really an estimate of individual variance based on the variance of sample averages.

5-8-11   Individual variance calculation

Individual variance can be estimated one other way, based on the actual individual variation. Using the formula for individual variance, the variance within the A_1 group can be calculated:

For the A_1 group

$$S_{y_2}^2 = 0.0570 \qquad \text{where } n = 4 \text{ and } v = 3$$

For the A_2 group

$$S_{y_2}^2 = 0.0630 \qquad \text{where } n = 4 \text{ and } v = 3$$

For the A_3 group

$$S_{y_2}^2 = 0.0370 \qquad \text{where } n = 4 \text{ and } v = 3$$

Each of these is an estimate of individual variance with 3 d.f. Which is the correct value? Obviously, all cannot be correct. Since all of this data is available, an obligation exists to use all of the information rather than a part of the information. A better estimate uses all of the information by averaging them:

$$S_{y_2}^2 = \frac{(0.0570 + 0.0630 + 0.0370)}{3} = 0.052$$

Now there are two independent estimates of individual variance; however, these estimates appear to be considerably different in this case:

$$1.008 >> 0.052$$

$$S_{y_1}^2 >> S_{y_2}^2$$

$S_{y_1}^2$ being determined from the variation of averages and $S_{y_2}^2$ being determined from the variation of individuals.

5-8-12 *F* test for variance comparison

Statistically, there is a tool which provides a decision at some confidence level as to whether these estimates are significantly different. This tool is called an F test, named after Sir Ronald Fisher, a British statistician, who invented the ANOVA method. The F test is simply a ratio of sample variances.

$$F = \frac{S_{y_1}^2}{S_{y_2}^2}$$

When this ratio becomes large enough, then the two sample variances are accepted as being unequal at some confidence level. F tables which list the required F ratios to achieve some confidence level are provided in the appendixes. To determine whether an F ratio of two sample variances is statistically large enough, three pieces of information are considered. One, the confidence level necessary; two, the degrees of freedom associated with the sample variance in the numerator; and three, the degrees of freedom associated with the sample variance in the denominator. Each combination of confidence, numerator degrees of freedom, and denominator degrees of freedom has an F ratio associated with it. $F_{\alpha;v_1;v_2}$ is the format for determining an explicit F value where

α = risk

Confidence = $1-$risk

$\quad v_1$ = degrees of freedom associated with the numerator

$\quad v_2$ = degrees of freedom associated with the denominator

In this example, an assumed confidence level of 90% is required, so the risk becomes 10%. The degrees of freedom for the numerator is 2 and for the denominator is 9. The necessary F ratio to look for in Table D-6 of Appendix D is

$$F_{.10;2;9}$$

Looking in the appropriate table, the 90% confidence table,

$$F_{.10;2;9} = 3.01$$

In the practice problem, the ratio of the two estimates of individual variance is:

$$F = \frac{1.008}{0.052} = 19.4$$

Comparing the F value of the data with the table,

$$F_{\text{data}} >> F_{.10;2;9}$$

Statistically, this means that with at least 90% confidence the two estimates of variance are believed to be unequal. What does this mean from a practical viewpoint?

The estimate of individual variance based on variation of averages is inappropriately too high when considering the actual individual variance. Averages are varying much more than would be expected from the individual variation that is present. Rather than believing that only one population is being sampled, it is believed that two or more having different averages are being sampled. The data from the practice problem in Sec. 5-8-8 is graphically displayed in Fig. 5-15. In another form the graph of the data would appear as in Fig. 5-16. In summary, at a very high confidence level (note: the F ratio from the practice problem exceeds the F ratio from the tables for 99% confi-

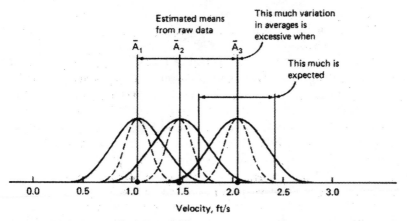

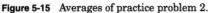

Figure 5-15 Averages of practice problem 2.

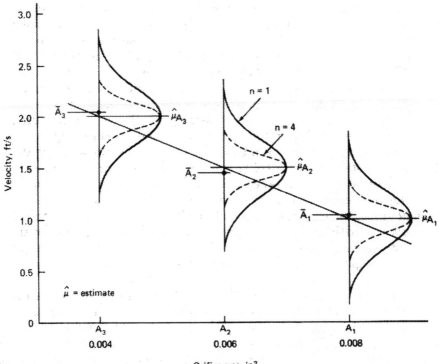

Figure 5-16 Averages versus orifice area.

dence level), the outlet orifice area affects flow velocity generated by the windshield washer pump.

As the F ratio is applied in the ANOVA case, the alpha risk is the chance of obtaining a ratio of at least the magnitude indicated in the table when pulling samples out of the same population (a group of items having the same mean and variance). Since the alpha risk is chosen as a small value, the experimenter would rather believe that two or more populations with different averages have been sampled when the F ratio does attain the specified magnitude. The odds are great against samples from one population attaining the F ratio; it is much more easily obtained when two or more populations are considered.

Confidence in the statistical sense, as it is used in this example, means there is some chance of error in the statement to which the confidence applies. For instance, the confidence that a number of 1, 2, 3, 4, or 5 could be rolled on a standard die is 5/6, or 83%. It is possible to roll a 6, so there is some risk of being wrong. When stating a confidence level for a factor, the experimenter is simply stacking the odds in their favor that the factor's levels really do have different average performances. A high confidence value may be chosen to reduce the risk of an error, alpha, but a larger actual difference in performance will be required to make the risk lower. In this situation the alpha risk is the chance that a factor really does not cause a difference in performance when the experimenter, based on the data at hand, believes that it does. The magnitude of the F ratio does not necessarily reflect the engineering importance of the factorial effect. There could exist a very high F ratio, if error variance were very small relative to the factor variance, with only a small difference (from an engineering viewpoint) in level averages. The confidence may be high, but the difference in level averages may not be practically useful to the experimenter.

The performance of each pump is best estimated to be linear over the range of orifice areas tested because the sample averages are within the expected distance of a set of averages falling on a straight line. A method to statistically evaluate nonlinear performance will be covered in Sec. 5-17. To obtain some particular flow velocity, the appropriate orifice area may be determined from the graph. In the previous graph the straight line predicts where the average of each population would fall, and the actual sample average from the experiment falls within the range expected due to the variation of individuals.

The interpretation of ANOVA results falls into two categories initially:

1. Factors which have an F ratio exceeding some criterion

2. Factors which have an F ratio less than some criterion

TABLE 5-14   Complete One-Way ANOVA Summary

Source	SS	v	V	F
A	2.0150	2	1.008	19.4#
e	0.4675	9	0.052	
T	2.4825	11		

+ at least 90% confidence
++ at least 95% confidence
# at least 99% confidence

The factors which have an F ratio larger than the criterion (F ratio from the tables) are believed to influence the average value for the population, and factors which have an F ratio less than the criterion are believed to have no effect on the average. A complete form of the ANOVA table would now appear as in Table 5-14. Method 2 and the format for the ANOVA summary shown in Table 5-14 will be the basic analytical technique used throughout the remainder of this text.

5-9   Two-Way ANOVA

Two-way ANOVA is the next highest order of ANOVA to review; there are two controlled parameters in this experimental situation. Method 2 will be used, but the graphical representations will be discontinued, although utilization is still possible.

5-9-1   Casting experiment analysis

During one seminar, a student proposed this experiment. He worked at an aluminum casting foundry which manufactured pistons for reciprocating engines. One problem existed at the end of the casting process, which was how to attain the proper hardness of the casting for a particular product. The hardness was measured on the Rockwell B scale. Engineers were interested in the effect of copper and magnesium content on casting hardness. According to specifications the copper content could be 3.5 to 4.5% and the magnesium content could be 1.2 to 1.8%. An experiment could be run to evaluate these factors and conditions simultaneously, using this symbology:

$$A = \% \text{ copper content} \quad A_1 = 3.5 \quad A_2 = 4.5$$

$$B = \% \text{ magnesium content} \quad B_1 = 1.2 \quad B_2 = 1.8$$

The experimental conditions can all be shown in Table 5-15. Four possible combinations exist: A_1B_1, A_1B_2, A_2B_1, and A_2B_2. Imagine that the four different mixes of metal constituents are prepared, castings poured, and resulting hardness measured. Two parts are randomly

TABLE 5-15   Two-Way Experimental Layout

	A_1	A_2
B_1		
B_2		

TABLE 5-16   Two-Way Experimental Data

	A_1	A_2
B_1	76, 78	73, 74
B_2	77, 78	79, 80

TABLE 5-17   Transformed Data for Two-Way ANOVA

	A_1	A_2
B_1	6, 8	3, 4
B_2	7, 8	9, 10

selected from each batch of castings and the hardness is measured. The results could quite possibly look like those in Table 5-16.

Remembering that the variation due to the mean will not be considered anyway, 70 points of hardness are subtracted from each value to simplify the discussion. Transformed results are shown in Table 5-17.

5-9-2   Sums of squares (two-way ANOVA)

In two-way ANOVA total variation may be decomposed into more components:

1. Variation due to factor A
2. Variation due to factor B
3. Variation due to the interaction of factors A and B
4. Variation due to error

An equation for total variation may be written

$$SS_T = SS_A + SS_B + SS_{A \times B} + SS_e$$

$A \times B$ represents the interaction of factors A and B. The interaction is the mutual effect of copper and magnesium in affecting casting hardness. If the effect on hardness of the percentage of copper depends on

TABLE 5-18   Two-Way Layout, Summarized Data

	A_1	A_2	Total	
B_1	6, 8	3, 4	21	
B_2	7, 8	9, 10	34	
Total	29	26	55	Grand Total

the percentage of magnesium, then an interaction is said to be present. This will be discussed in more detail in Sec. 5-9-4.

Some preliminary calculations will speed the ANOVA for this experiment, as summarized in Table 5-18.

$$A_1 = 29 \qquad B_1 = 21 \qquad T = 55$$
$$A_2 = 26 \qquad B_2 = 34$$
$$n_{A_1} = 4 \qquad n_{B_1} = 4 \qquad N = 8$$
$$n_{A_2} = 4 \qquad n_{B_2} = 4$$

The total variation is

$$SS_T = \left[\sum_{i=1}^{N} y_i^2 \right] - \frac{T^2}{N}$$

$$SS_T = 6^2 + 8^2 + 3^2 + \ldots + 10^2 - \left(\frac{55^2}{8} \right) = 40.875$$

The variation due to factor A can be calculated two different, but equivalent, ways. The general formula for any number of levels of factor A is

$$SS_A = \left[\sum_{i=1}^{k_A} \left(\frac{A_i^2}{n_{A_i}} \right) \right] - \frac{T^2}{N}$$

$$= \frac{A_1^2}{n_{A_1}} + \frac{A_2^2}{n_{A_2}} + \ldots + \frac{A_k^2}{n_{A_k}} - \frac{T^2}{N}$$

In a two-level situation the formula may be written

$$= \frac{A_1^2}{n_{A_1}} + \frac{A_2^2}{n_{A_2}} - \frac{T^2}{N}$$

$$= \frac{29^2}{4} + \frac{26^2}{4} - \frac{55^2}{8} = 1.125$$

This is an opportune time to point out a mathematical check of the SS_A calculation. Note that

$$29 + 26 = 55 \qquad \text{and} \qquad 4 + 4 = 8$$

The sum of the numerators of the plus terms (neglecting the square) must equal the numerator of the negative term. The sum of the denominators of the plus terms must equal the denominator of the negative term. If these conditions are not met, then the SS_A calculation will be wrong. For a two-level experiment when sample sizes are equal, this equation can be simplified to this special formula (see Proof 2 in App. E):

$$SS_A = \frac{(A_1 - A_2)^2}{N}$$

$$= \frac{(29 - 26)^2}{8} = \frac{3^2}{8} = 1.125$$

Similarly, the variation due to factor B is

$$SS_B = \frac{(B_1 - B_2)^2}{N}$$

$$= \frac{(21 - 34)^2}{8} = 21.125$$

To calculate the variation due to the interaction of factors A and B, the same concept of squaring the magnitude of the line segments due to the factor A and B combinations is applied but will not be graphically shown. To simply show the calculation method will suffice at this point. When the variation due to factor A was calculated, the data was organized in the fashion of Table 5-19. When variation due to factor B was calculated, the data was organized as in Table 5-20. To

TABLE 5-19   Data Summarized over Factor A

	A_1	A_2	
4 data points make up this total	29	26	4 data points make up this total

TABLE 5-20   Data Summarized over Factor B

B_1	21	4 data points make up each
B_2	34	of these totals

TABLE 5-21   Data Summarized over Interaction $A \times B$

	A_1	A_2	
B_1	14	7	2 data points make up
B_2	15	19	each of these totals

calculate the variation due to the interaction of factors A and B, the data must be organized in the manner shown in Table 5-21. This data is organized into all the possible factor A and B combinations and the data summed for each combination. Discussed subsequently are two methods for calculating the variation due to an interaction. The first is a general method for any number of levels of either factor and the second is a specific method for only two-level factors.

General formula. Let $(A{\times}B)_i$ represent the sum of data under the ith condition of the combinations of factor A and B. Also, let c represent the number of possible combinations of the interacting factors and $n_{(A{\times}B)_i}$ the number of data points under this condition. Then

$$SS_{A{\times}B} = \left[\sum_{i=1}^{c} \left(\frac{(A{\times}B)_i^2}{n_{A{\times}B_i}} \right) \right] - \frac{T^2}{N} - SS_A - SS_B$$

So for the example problem,

$$A_1B_1 = (A \times B)_1 = 14$$

$$A_2B_1 = (A \times B)_2 = 7$$

$$A_1B_2 = (A \times B)_3 = 15$$

$$A_2B_2 = (A \times B)_4 = 19$$

$$c = 4$$

$$SS_{A{\times}B} = \frac{14^2}{2} + \frac{7^2}{2} + \frac{15^2}{2} + \frac{19^2}{2} - \frac{55^2}{8} - 1.125 - 21.125$$

$$SS_{A{\times}B} = 15.125$$

Note that when the various combinations are summed, squared, and divided by the number of data points for that combination, the subsequent value also includes the factor main effects which must be subtracted. When interaction effects are calculated using the general formula, all lower-order interactions and factor effects must be subtracted.

Specific formula. In the test data, summations were made in two directions to assess the main effects of factors A and B, vertically to obtain the effect of factor A, and horizontally to obtain the effect of factor B. In a two-factor, two-level experiment, one other summation possibility exists which assesses the interaction effect, that is, diagonally. $A_1B_1 + A_2B_2 = \underline{A \times B}_1$ which represents one diagonal sum and $A_1B_2 + A_2B_1 = \underline{A \times B}_2$ which represents the other diagonal sum. Since one comparison of two groups is made, the specific formula may be applied:

$$SS_{A \times B} = \frac{(A \times B_1 - A \times B_2)^2}{N}$$

$$SS_{A \times B} = \frac{(33 - 22)^2}{8} = 15.125$$

Note that this method includes the interactive effect only; the lower-order interactions and factor effects need not be subtracted.

The easiest way to calculate error variation in this case and also more complex ANOVAs is to use this technique. Whatever variation is left over from that which is accountable is attributed to error, where

$$SS_T = SS_A + SS_B + SS_{A \times B} + SS_e$$

Then

$$SS_e = SS_T - SS_A - SS_B - SS_{A \times B}$$

$$SS_e = 40.875 - 1.125 - 21.125 - 15.125 = 3.500$$

5-9-3   Degrees of freedom (two-way ANOVA)

The degrees of freedom for all items except the interaction are identical to the Method 2 calculations.

Total degrees of freedom:

$$v_T = N - 1 = 8 - 1 = 7$$

also

$$v_T = v_A + v_B + v_{A \times B} + v_e$$

Factor A degrees of freedom:

$$v_A = k_A - 1 = 2 - 1 = 1$$

Factor B degrees of freedom:

$$v_B = k_B - 1 = 2 - 1 = 1$$

Interaction $(A \times B)$ degrees of freedom:

$$v_{A \times B} = (v_A)(v_B) = 1 \times 1 = 1$$

Error degrees of freedom:

$$v_e = v_T - v_A - v_B - v_{A \times B}$$

$$v_e = 7 - 1 - 1 - 1 = 4$$

The 4 d.f. for error are readily apparent in this situation by referring to Table 5-17. One fair comparison between two data points within each treatment condition is available.

5-9-4 ANOVA summary table (two-way ANOVA)

The summary of the ANOVA results is shown in Table 5-22. The ANOVA results indicate that copper content by itself has no effect on the resultant casting hardness, magnesium content by itself has a substantial effect (largest SS) on hardness, and the interaction of copper and magnesium content plays a substantial part in determining hardness. A plot of the test data shows this in Fig. 5-17.

In this plot, there exist nonparallel lines which indicate the presence of an interaction. The factor A effect depends on the level of factor B and vice versa. If the lines were parallel, there would be no interaction (the factor A effect would be the same regardless of the level of factor B). The average of all results under A_1 and all results under A_2 are not very different and the average of all results under B_1 and all results under B_2 are different. The repeated data points at each interaction combination imply there is some error variation around the average for that combination. Note that in this case the conditions $A_1 B_1$ and $A_1 B_2$ do not appear to be different. Later, a method will be shown which will aid in making the decision of whether or not this is true.

Geometrically, there is some information available from the graph that may be useful, as seen in Fig. 5-18. The relative magnitudes of the various effects can be seen graphically. The B effect is the largest, the $A \times B$ effect next largest, and the A effect is very small.

TABLE 5-22   ANOVA of Casting Experiment

Source	SS	v	V	F
A	1.125	1	1.125	1.29
B	21.125	1	21.125	24.14#
$A \times B$	15.125	1	15.125	17.29++
e	3.500	4	0.875	
T	40.875	7		

+ at least 90% confidence
++ at least 95% confidence
# at least 99% confidence

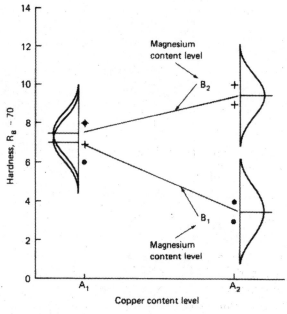

Figure 5-17   Copper-magnesium interaction.

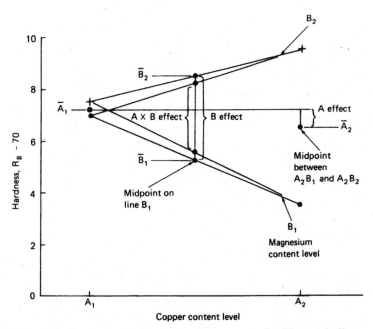

Figure 5-18   Copper-magnesium interaction; magnitude of factorial effects.

More interpretation of this kind of information will be discussed in later chapters.

5-10 Three-Way ANOVA

One final ANOVA method will be discussed which will demonstrate some mathematical complications in ANOVA. Three-way ANOVA entails three controlled factors in an experiment.

Dispensing with any assumed experimental situation, the arrangement for the experiment would appear like Table 5-23. Here is a case where factor B has three levels and other factors two levels. The test data could be as in Table 5-24.

5-10-1 Sums of squares (three-way ANOVA)

An equation for total variation may be written in this case:

$$SS_T = SS_A + SS_B + SS_C + SS_{A\times B} + SS_{A\times C} + SS_{B\times C} + SS_{A\times B\times C} + SS_e$$

Also, the total variation may be calculated

TABLE 5-23   Three-Way Experiment Layout

	A_1	A_2	
B_1			C_1
			C_2
B_2			C_1
			C_2
B_3			C_1
			C_2

TABLE 5-24   Data for Three-Way Experiment

	A_1	A_2	
B_1	1 3	7 9	C_1
	3 5	8 12	C_2
B_2	4 8	10 12	C_1
	10 14	12 16	C_2
B_3	6 8	11 13	C_1
	13 15	15 17	C_2

TABLE 5-25   Factor *A* Effect

A_1	A_2
90	142

$$SS_T = \left[\sum_{i=1}^{N} y_i^2 \right] - \frac{T^2}{N}$$

$$= 1^2 + 3^2 + 7^2 + \ldots + 15^2 + 17^2 - \frac{232^2}{24} = 457.333$$

To calculate the variation due to factor *A*, the data is first summed as shown in Table 5-25.

$$SS_A = \frac{(A_1 - A_2)^2}{N} = \frac{(90 - 142)^2}{24} = 112.667$$

The data can be organized as in Table 5-26, and the variation due to factor *B* determined:

$$SS_B = \frac{B_1^2}{n_{B_1}} + \frac{B_2^2}{n_{B_2}} + \frac{B_3^2}{n_{B_3}} - \frac{T^2}{N}$$

$$SS_B = \frac{48^2}{8} + \frac{86^2}{8} + \frac{98^2}{8} - \frac{232^2}{24} = 170.333$$

When the data is organized as in Table 5-27, the variation due to factor *C* may be calculated:

$$SS_C = \frac{(C_1 - C_2)^2}{N} = \frac{(92 - 140)^2}{24} = 96.000$$

To calculate the interaction the data must be arranged into all the

TABLE 5-26   Factor *B* Effect

B_1	48
B_2	86
B_3	98

TABLE 5-27   Factor *C* Effect

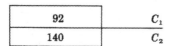

92	C_1
140	C_2

TABLE 5-28 $A \times B$ Interaction Effect

	A_1	A_2
B_1	12	36
B_2	36	50
B_3	42	56

possible combinations of factors A and B and summed within those combinations as shown in Table 5-28.

$$SS_{A \times B} = \left[\sum_{i=1}^{c} \left(\frac{(A \times B)_i^2}{n_{A \times B_i}} \right) \right] - \frac{T^2}{N} - SS_A - SS_B$$

$$SS_{A \times B} = \frac{12^2 + 36^2 + 36^2 + 50^2 + 42^2 + 56^2}{4} - \frac{232^2}{24} - 112.667 - 170.333$$

$$SS_{A \times B} = 8.333$$

The variation due to the interaction of factors A and C may be determined after the data is organized into combinations of factor A and factor C shown in Table 5-29.

$$SS_{A \times C} = \frac{(122 - 110)^2}{24} = 6.000$$

Variation due to the interaction of factors B and C is calculated similarly when the six combinations of factor B and C are arranged in Table 5-30.

$$SS_{B \times C} = \frac{20^2 + 28^2 + 34^2 + 52^2 + 38^2 + 60^2}{4} - \frac{232^2}{24} - 170.333 - 96.000$$

$$SS_{B \times C} = 13.000$$

To determine the variation due to the interaction of factors A, B, and C, the data must be organized into all the possible factor A, B, and C combinations and summed for that treatment condition like the arrangement in Table 5-31.

TABLE 5-29 $A \times C$ Interaction Effect

	A_1	A_2	
	30	62	C_1
	60	80	C_2
122			110

TABLE 5-30 $B \times C$ Interaction Effect

B_1	20	C_1
	28	C_2
B_2	34	C_1
	52	C_2
B_3	38	C_1
	60	C_2

TABLE 5-31 $A \times B \times C$ Interaction Effect

	A_1	A_2	
B_1	4	16	C_1
	8	20	C_2
B_2	12	22	C_1
	24	28	C_2
B_3	14	24	C_1
	28	32	C_2

$$SS_{A \times B \times C} = \left[\sum_{i=1}^{c} \left(\frac{(A \times B \times C)_i^2}{n_{(A \times B \times C)_i}} \right) \right] - \frac{T^2}{N} - SS_A - SS_B - SS_C - SS_{A \times B}$$
$$- SS_{A \times C} - SS_{B \times C}$$

Note that all lower-order interactions and factor effects must be subtracted from the summation.

$$SS_{A \times B \times C} = \frac{4^2 + 16^2 + 8^2 + \ldots + 32^2}{2} - \frac{232^2}{24} - 112.667$$
$$- 170.333 - 96.000 - 8.333 - 6.000 - 13.000 = 3.000$$

Error can be determined by finding the remainder of the total variation left from the known sources:

$$SS_e = SS_T - SS_A - SS_B - SS_C - SS_{A \times B} - SS_{A \times C} - SS_{B \times C} - SS_{A \times B \times C}$$

$$SS_e = 457.333 - 112.667 - 170.333 - 96.000 - 8.333 - 6.000$$
$$- 13.000 - 3.000$$
$$SS_e = 48.000$$

5-10-2   Degrees of freedom
(three-way ANOVA)

The degrees of freedom for the various factors and interactions are

$$v_T = v_A + v_B + v_C + v_{A \times B} + v_{A \times C} + v_{B \times C} + v_{A \times B \times C}$$
$$v_T = N - 1 = 24 - 1 = 23$$
$$v_A = k_A - 1 = 2 - 1 = 1$$
$$v_B = k_B - 1 = 3 - 1 = 2$$
$$v_C = k_C - 1 = 2 - 1 = 1$$
$$v_{A \times B} = (v_A)(v_B) = (1)(2) = 2$$
$$v_{A \times C} = (v_A)(v_C) = (1)(1) = 1$$
$$v_{B \times C} = (v_B)(v_C) = (1)(2) = 2$$
$$v_{A \times B \times C} = (v_A)(v_B)(v_C) = (1)(2)(1) = 2$$
$$v_e = v_T - v_A - v_B - v_C - v_{A \times B} - v_{A \times C} - v_{B \times C} - v_{A \times B \times C}$$
$$v_e = 23 - 1 - 2 - 1 - 2 - 1 - 2 - 2 = 12$$

5-10-3   ANOVA summary table
(three-way ANOVA)

The ANOVA summary is shown in Table 5-32. A procedure to utilize

TABLE 5-32   Three-Way ANOVA Summary

Source	SS	v	V	F
A	112.667	1	112.667	28.17#
B	170.333	2	85.167	21.29#
C	96.000	1	96.000	24.00#
$A \times B$	8.333	2	4.167	1.04
$A \times C$	6.000	1	6.000	1.50
$B \times C$	13.000	2	6.500	1.63
$A \times B \times C$	3.000	2	1.500	0.38
e	48.000	12	4.000	.
T	457.333	23		

+ at least 90% confidence
++ at least 95% confidence
# at least 99% confidence

information associated with insignificant factors and/or interactions will be discussed in later sections.

At this point, main effects A, B, and C all appear to be significant in affecting the average in this experiment. Plots should be made and interpreted accordingly.

5-11   Critique of the F Test

There are limitations in this method, of which the reader should be aware. Referencing the practice problem in Sec. 5-8-8, the error variance of 0.052 was obtained by averaging the sample variances from each of the three controlled groups of the levels of factor A. This happens automatically in ANOVA because the sums of squares for all estimates of error are pooled together and divided by the total degrees of freedom for error. A basic assumption of ANOVA is that error variance is equal for all treatment conditions (combinations of the various levels of the various factors); however, this may not be true. Because of the automatic averaging, an opportunity to reduce variation by controlling levels of design parameters may go unrecognized. Because of the loss function, opportunities to reduce variation should be sought out and utilized. Therefore, the F test will be used as a reference decision-making tool and other methods will be utilized to detect a reduction in variation.

Another limitation of the F test is that only the alpha risk is addressed, which is the risk of saying a factor affects the average performance when in fact it does not. Another risk which is the opposite side of the same coin is the beta risk. The beta risk is the risk of saying a factor has no effect on the average performance when in fact it does. Since the beta risk is not assessed, the experimenter may not know what chance there is of missing a factor that does make a difference. The alpha and beta risks will be discussed more in Sec. 5-13-4.

5-12   Two-Way ANOVA Applied to L4 OA

Two-way ANOVA can be adapted to the L4 OA. Using the same die-cast piston example as in Sec. 5-9-1, factor A can be assigned to column 1 and factor B can be assigned to column 2. The first trial then represents the A_1B_1 condition, which has results of 6 and 8. Trial 2 represents the A_1B_2 condition, with results of 7 and 8. The remaining two trials follow the same pattern which makes the entire experiment look like Table 5-33.

Typically, OAs are analyzed in the same manner as other structured experiments. Recall the ANOVA summary in Sec. 5-9-4, for the example in Sec. 5-9-1.

TABLE 5-33 L4 OA with Casting Data

	Factors				
	A	B			
	Column no.				
Trial no.	1	2	3	y data	(R_B-70)
1	1	1	1	6	8
2	1	2	2	7	8
3	2	1	2	3	4
4	2	2	1	9	10

$$SS_T = 40.875$$
$$SS_A = 1.125$$
$$SS_B = 21.125$$
$$SS_{A \times B} = 15.125 \text{ and}$$
$$SS_e = 3.500$$

The ANOVA for an OA is conducted by calculating the sums of squares for each column. The formula for SS_A is the same as in Sec. 5-9-2. The sums of squares for factor A, column 1, is

$$SS_A = \frac{(A_1 - A_2)^2}{N}$$

A_1 and A_2 are the sums of the data associated with the first and second levels of factor A, respectively. The sums of squares calculation uses the same piece of information as the difference row of the column effects method. The column effects method is really an early stopping point in the ANOVA method. Decisions are based on the same numerical difference; in the column effects method the decision is subjective, and in ANOVA the decision is objective. The level sums are

$$A_1 = 6 + 8 + 7 + 8 = 29$$
$$A_2 = 3 + 4 + 9 + 10 = 26$$
$$SS_A = \frac{(29 - 26)^2}{8} = 1.125$$

The sums of squares for factor B, column 2, is calculated in the same manner:

$$SS_B = \frac{(B_1 - B_2)^2}{N}$$

$$B_1 = 6 + 8 + 3 + 4 = 21$$
$$B_2 = 7 + 8 + 9 + 10 = 34$$
$$SS_B = \frac{(21 - 34)^2}{8} = 21.125$$

Note that the sums of squares for factors A and B are identical to those in Sec. 5-9-2. The sums of squares for column 3 is

$$SS_3 = \frac{(3_1 - 3_2)^2}{N} = \frac{(33 - 22)^2}{8} = 15.125$$

Note that this value is equal to the $SS_{A \times B}$, which is not coincidental but is a mathematical property of the OA. The calculation is a demonstration, not a proof, that the third column represents the interaction of the factors assigned to the first and second columns. The method to determine interaction columns was covered in Chap. 3, Sec. 3-3-5.

This particular L4 example is similar to the two-way ANOVA example, since each trial has two test results which provide an estimate of error variance with 4 degrees of freedom. The sums of squares due to error can be calculated:

$$SS_e = SS_T - SS_A - SS_B - SS_{A \times B}$$
$$SS_e = 40.875 - 1.125 - 21.125 - 15.125 = 3.500$$

The L4 OA having two factors assigned to it is equivalent to a full-factorial experiment and the ANOVA is equivalent to a two-way ANOVA because certain columns in OAs represent the interaction of two other columns.

5-13   Two-Way ANOVA Applied to L8 OA

This simple example is intended to demonstrate another basic property of OAs, which is that the total variation can be accounted for by summing the variation from all columns. Another technique used in OA analysis is the concept of pooling together all of the estimates of error variance.

5-13-1   ANOVA for an L8 OA

One commonly used Taguchi OA is an L8, shown earlier in Table 3-5. This OA also can be used to display the experimental structure used in the example in Sec. 5-9-1. Factor A can be assigned to column 1 and factor B assigned to column 2 of the L8 OA. The first two trials of the OA now represent the A_1B_1 condition, which has corresponding results of 6 and 8 in the example. Trials 3 and 4 represent the A_1B_2 condition, which has results corresponding to 7 and 8. The complete

TABLE 5-34   Two-Factor L8 OA with Data

Trial no.	A	B	A×B					y data
			Column no.					$(R_B - 70)$
	1	2	3	4	5	6	7	
1	1	1	1	1	1	1	1	6
2	1	1	1	2	2	2	2	8
3	1	2	2	1	1	2	2	7
4	1	2	2	2	2	1	1	8
5	2	1	2	1	2	1	2	3
6	2	1	2	2	1	2	1	4
7	2	2	1	1	2	2	1	9
8	2	2	1	2	1	1	2	10

OA and results are shown in Table 5-34.

Again, the ANOVA for an OA is conducted by calculating the sums of squares for each column. The formula for SS_A is the same as Sec. 5-9-2 and the L4 example in Sec. 5-12. The sums of squares for factor A, column 1, is

$$SS_A = \frac{(A_1 - A_2)^2}{N}$$

A_1 and A_2 are the sums of the data associated with the first and second levels of factor A, respectively,

$$A_1 = 6 + 8 + 7 + 8 = 29$$
$$A_2 = 3 + 4 + 9 + 10 = 26$$
$$SS_A = \frac{(29 - 26)^2}{8} = 1.125$$

The sums of squares for factor B, column 2, is calculated in the same manner

$$SS_B = \frac{(B_1 - B_2)^2}{N}$$

$$B_1 = 6 + 8 + 3 + 4 = 21$$
$$B_2 = 7 + 8 + 9 + 10 = 34$$
$$SS_B = \frac{(21 - 34)^2}{8} = 21.125$$

The sums of squares for column $A \times B$ is

$$SS_{A \times B} = \frac{(3_1 - 3_2)^2}{N} = \frac{(33 - 22)^2}{8} = 15.125$$

Note that the sums of squares for factors A and B and interaction $A \times B$ are identical to the example in Secs. 5-9-2 and 5-12. Continuing with the sums of squares calculations,

$$SS_4 = \frac{(4_1 - 4_2)^2}{N} = \frac{(25 - 30)^2}{8} = 3.125$$

$$SS_5 = \frac{(5_1 - 5_2)^2}{N} = \frac{(27 - 28)^2}{8} = 0.125$$

$$SS_6 = \frac{(6_1 - 6_2)^2}{N} = \frac{(27 - 28)^2}{8} = 0.125$$

$$SS_7 = \frac{(7_1 - 7_2)^2}{N} = \frac{(27 - 28)^2}{8} = 0.125$$

$$SS_e = SS_4 + SS_5 + SS_6 + SS_7 = 3.500$$

Note that the total of the sums of squares for the unassigned columns is equal to error sums of squares. The unassigned, truly empty columns in an OA represent an estimate of error variation. Also, as in Sec. 5-4-1, there are 4 degrees of freedom (four columns) associated with error variation. Here the difference of the particular array selected for the experiment changes the analysis approach slightly. The L4 has 2 data points per trial and the L8 1 data point per trial. The error variance of the L4 must come from the repetitions in each trial, but the error variance in the L8 must come from the columns, since there are no repetitions within trials.

Note also that the total of all the column sums of squares for the L8 OA is equal to SS_T for the 8 data points. This is a demonstration of the property of the total sums of squares being contained within the columns of an OA.

$$SS_T = SS_{columns}$$

In summary, we see the sums of squares and the degrees of freedom associated with each component of variation correlate exactly, whether a 2×2 factorial arrangement or a two-factor, two-level OA. In this case the ANOVA for the OA is identical to a two-way ANOVA, which means that this OA example is really a full-factorial experiment.

5-13-2   Column estimates of error variance

In the previous example, the unassigned columns were shown to estimate error variance when there was only one test result per trial. This approach of using columns to estimate error variance will be used even if all columns have factors assigned to them.

Recalling the ANOVA chapter, the variance due to a factor is really an estimate of the variance of individual data points (same as error variance) based on the variance of the sample averages of that factor. When assigned columns are used, it is obvious that the variation of the average of the data associated with level 1 and level 2 can only be attributed to unknown and uncontrolled factors. In fact, the level 1s and levels 2s only represent groups of data points; when no factor or interaction is present in a column, the 1s and 2s are meaningless. Hopefully, this variation will be small; excessive variation will indicate that a potentially important factor has been excluded from the experiment.

When factors are assigned to all columns, error variance may still be estimated. Some factors assigned to an experiment will not be significant at all, even though they were thought to be so before experimentation. This would be equivalent to saying the color of a car can affect fuel economy and assigning two different colors to a column. It is very likely that this column will have a small sums of squares because it will really be estimating error variance rather than any true color effect.

When column effects turn out small in an OA then there are several possibilities:

1. No assigned factors or interactions, true error estimate

2. No significant factor and/or interaction effect

3. Very small factor and/or interaction effects

4. Canceling factor and/or interaction effects

With a low-resolution OA there is no proof that any one of the last three conditions is true or untrue, but either the second or third possibility will be accepted as true when column effects are small. From a practical point of view, there is no difference between situation 2 or 3.

A fully saturated OA will depend on some column effects turning out small relative to others and using the smaller ones as estimates of error variance.

5-13-3 Pooling estimates of error variance

In the ANOVA of Sec. 5-13-1, there were four unassigned columns

---

*G. E. P. Box, W. G. Hunter, and J. S. Hunter, *Statistics for Experimenters*. Wiley, New York, 1978, pp. 374–376.

having 4 degrees of freedom, one for each column, which provided estimates of error variance. A better estimate was the combination of all four column effects for one overall estimate of error variance with 4 degrees of freedom. The combining of column effects to better estimate error variance is referred to as *pooling*.

Two pooling strategies exist:

1. Pooling up (Taguchi)

2. Pooling down*

The pooling-up strategy entails F-testing the smallest column effect against the next larger one to see if significance exists. If no significant F ratio exists, then these two effects are pooled together to test the next larger column effect until some significant F ratio exists.

The pooling-down strategy entails pooling all but the largest column effect and F-testing the largest against the remainder pooled together. If that column effect is significant, then the next largest is removed from the pool and those two column effects are F-tested against all others pooled until some insignificant F ratio is obtained.

The pooling-up strategy will tend to maximize the number of columns judged to be significant, and the pooling-down strategy will tend to minimize the number of significant columns. What is the penalty of judging too many columns as being significant or too few?

5-13-4 Alpha and beta mistakes

In Sec. 2-1 there was reference to improving the quality of the decision of whether to use a new design, process, method, etc., or not to use a new design based on test data. When making this decision, there are four possible outcomes, as shown in Table 5-35. When using the pooling-up strategy and judging many columns to be significant, the decision will be to use these factors for further experimentation and per-

TABLE 5-35   Decision Risks

		The truth about the product	
		There is no improvement	There is some improvement
Decision based on test data	Do not use new design	OK	Beta mistake
	Do use new design	Alpha mistake	OK

haps product or process design. The tendency will be to make the alpha mistake more often, thinking that some factor will cause an improvement, when, in truth, that factor will not help. When using the pooling-down strategy and judging few columns to be significant, the decision will be to ignore many factors and use only a few for further experimentation and perhaps product or process design. The tendency will be to make the beta mistake more often, thinking that some factor makes no improvement, when, in truth, that factor will help.

From the customer's viewpoint, it is much more serious to make a beta mistake and not take advantage of some amount of improvement offered by a factor. Also, once a factor has been judged to be insignificant, that factor will probably not be included in further rounds of experimentation and the beta mistake will never be exposed. However, if an alpha mistake is made, that factor will be included in further experimentation and the alpha mistake will potentially be exposed. Since it is impossible to make both the alpha and beta mistakes simultaneously, the pooling-up strategy should be used, which will tend to prevent the beta mistake of ignoring helpful factors. There are books available which explain the alpha-beta risk and the economic consequences of those mistakes.*

5-13-5 Example of pooling error variance estimates

TABLE 5-36 Pooling of Error Variance

Source	SS	v	V	F
A*	1.125	1	1.125	
B	21.125	1	21.125	22.83#
$A \times B$	15.125	1	15.125	16.35++
Col 4*	3.125	1	3.125	
Col 5*	0.125	1	0.125	
Col 6*	0.125	1	0.125	
Col 7*	0.125	1	0.125	
T	40.875	7		
e pooled	4.625	5	.925	

*Indicates those items that were combined to generate the pooled error estimate.
  + at least 90% confidence
  ++ at least 95% confidence
  # at least 99% confidence

---

*W. J. Diamond, *Practical Experiment Designs for Engineers and Scientists*. Lifetime Learning Publications, Belmont, Calif., 1981, chap. 2.

TABLE 5-37   Pooling Error Variance ANOVA Summary Table

Source	SS	v	V	F
B	21.125	1	21.125	22.83#
A×B	15.125	1	15.125	16.35++
e_p	4.625	5	0.925	
T	40.875	7		

+ at least 90% confidence
++ at least 95% confidence
# at least 99% confidence

The ANOVA table for the experimental example, Sec. 5-13-1, that has been used predominantly throughout this section would appear as in Table 5-36. In this situation, five smaller column effects have been pooled to form an estimate of error variance with 5 degrees of freedom associated with that estimate. As a rule of thumb, pooling up to half of the degrees of freedom in an experiment is advisable. Here that rule was exceeded slightly, because two of the column effects were substantially larger than the others. The ANOVA summary table could be rewritten to recognize the pooling as shown in Table 5-37.

5-14   Percent Contribution

The portion of the total variation observed in an experiment attributed to each significant factor and/or interaction is reflected in the percent contribution P. The percent contribution is a function of the sums of squares for each significant item. The percent contribution indicates the relative power of a factor and/or interaction to reduce variation. If the factor and/or interaction levels were controlled precisely, then the total variation could be reduced by the amount indicated by the percent contribution.

When using an experiment to resolve a production problem, the total variation observed in an experiment should represent a large portion of the variation observed in production (see Sec. 4-3-4). If this is true, then the factors and/or interactions with substantial percent contributions are the most important items to control. This provides a basis for the experimenter to justify the additional costs of control that may be present. The loss function will play a part in these cost decisions, as well.

---

*C. R. Hicks, *Fundamental Concepts in the Design of Experiments*. New York, Holt, Rinehart, & Winston, 1982.

†P. W. M. John, *Statistical Design and Analysis of Experiments*. New York, Macmillan, 1971.

5-14-1   Calculating percent contribution

Although it has not been mentioned in the ANOVA sections, the variance due to a factor or interaction includes some amount due to error.*† An equation that states this for factor A, for example, is

$$V_A = V_A' + V_e$$

V_A' is the expected amount of variance due solely to factor A. Solving for $V_{A'}$,

$$V_A' = V_A - V_e \tag{5-3}$$

Recall that the definition of variance for factor A is

$$V_A = \frac{SS_A}{v_A}$$

Then

$$V_A' = \frac{SS_A'}{v_A}$$

Substituting into Eq. (5-3),

$$\frac{SS_A'}{v_A} = \frac{SS_A}{v_A} - V_e$$

Solving for SS_A'

$$SS_A' = SS_A - (V_e)(v_A)$$

SS_A' is the expected sum of squares due to factor A, and the percent P of the contribution to the total variation can now be calculated.

$$P = \left[\frac{SS_A'}{SS_T} \right] \times 100$$

This example used factor A, but any factor or interaction can be substituted as long as the appropriate sums of squares and associated degrees of freedom are used. One more column, P, can be added to complete an ANOVA table using Taguchi techniques. Since some portion of the sums of squares for a factor and/or interaction was subtracted out because of error, this amount must be added to the error sum of squares in order that the total sum of squares is unchanged. Since the total percent contribution must add up to 100 percent, the error contribution can be calculated by subtracting all the accountable sources from 100 percent.

5-14-2   Interpreting percent contribution

The percent contribution due to error provides an estimate of the adequacy of the experiment. If the percent contribution due to error (unknown and uncontrolled factors) is low (15% or less), then it is assumed that no important factors were omitted from the experiment. If it is a high value (50% or more), then some important factors were definitely omitted, conditions were not precisely controlled, or measurement error was excessive. The percent contribution of error is a measure of how much work is left to do or how much opportunity still exists. If the percent contribution for error is small and many factors were considered in the original screening experiment, then the opportunity for further improvement is not very great. However, if the percent contribution due to error is high, there is a good opportunity for further improvement and more experimentation may prove beneficial.

5-14-3 Casting experiment

The pooled ANOVA summary for the casting experiment shown in the previous table would now appear as in Table 5-38. Note that the percent contribution indicates that factor B, percent magnesium, all by itself contributes the most toward the variation observed in the experiment: almost 50 percent. The percent copper–percent magnesium interaction contributes over a third of the total variation observed. Even though factor A, percent copper, all by itself was not statistically significant, the fact that it is part of the interaction makes factor A important. The percent contribution of the interaction can be controlled only if both factors of the interaction are controlled.

The relative amounts of the percent contributions for the factors and interactions are in direct agreement with the magnitudes of the effects shown in the interaction plot in Fig. 5-18.

5-15 OA Column Confounding

Recalling from Sec. 5-9, "Two-Way ANOVA," the factor and interaction effects are estimated in the following manner. The data is summed

TABLE 5-38 Percent Contribution ANOVA Summary Table

Source	SS	v	V	F	P
B	21.125	1	21.125	22.83#	49.42
$A \times B$	15.125	1	15.125	16.35++	34.74
e_p	4.625	5	0.925		15.84
T	40.875	7			100.00

+ at least 90% confidence
++ at least 95% confidence
# at least 99% confidence

TABLE 5-39   Resolution 2 Experiment Confounding Demonstration

			Factors and interactions			
		$A \times B$		$A \times C$	$A \times D$	
A	B	$C \times D$	C	$B \times D$	$B \times C$	D
			Column no.			
Trial no.	1	2	3	4	5	6	7

Trial no.	1	2	3	4	5	6	7
1	1	1	1	1	1	1	1
2	1	1	1	2	2	2	2
3	1	2	2	1	1	2	2
4	1	2	2	2	2	1	1
5	2	1	2	1	2	1	2
6	2	1	2	2	1	2	1
7	2	2	1	1	2	2	1
8	2	2	1	2	1	1	2

horizontally, vertically, and diagonally to obtain the factor A effect, factor B effect, and the $A \times B$ interaction effect, respectively. When applied to an OA, treatment conditions A_1B_1 and A_2B_2 represent interaction condition $\underline{A \times B_1}$; A_1B_2 and A_2B_1 represent $\underline{A \times B_2}$. However, if a resolution 2 experiment is chosen, as shown in Table 5-39, there is confounding of interactions present. If the same representation is used for factors C and D, then C_1D_1 and C_2D_2 are represented by $\underline{C \times D_1}$; C_1D_2 and C_2D_1 are represented by $\underline{C \times D_2}$. One can observe that the third column follows the same pattern whether using the $A \times B$ interaction or the $C \times D$ interaction, thus the confounding of the two interactions and the analytical impossibility of separating the interaction effects from each other.

From an interpretational standpoint, however, if a column effect is large and there are only interactions located in that column (no factors assigned), then the interaction assumed to cause the effect will be the one that has factors which have a significant effect. If factors A and B have an effect and the third column appears to have an effect as well, then the interaction assumed to be causing that effect will be the $A \times B$ interaction and not $C \times D$.

5-16   Example Experiments

The two example experiments that have been discussed throughout the text provide the opportunity to apply many of the aspects of ANOVA, percent contribution, and confounding covered in this chapter.

TABLE 5-40   Water Pump ANOVA Summary (Unpooled)

Source	SS	v	V	F	P
Cover design	0.125	1	0.125	—	0.66
Gasket design	15.125	1	15.125	—	80.13
Front bolt torque	0.125	1	0.125	—	0.66
Sealant	0.125	1	0.125	—	0.66
Pump finish	3.125	1	3.125	—	16.57
Back bolt torque	0.125	1	0.125 °	—	0.66
Torque sequence	0.125	1	0.125	—	0.66
Total	18.875	7	—	—	100.00

TABLE 5-41   Water Pump ANOVA Summary (Pooled)

Source	SS	v	V	F	P
Gasket design	15.125	1	15.125	121#	79.47
Pump finish	3.125	1	3.125	25#	15.89
e_p	0.625	5	0.125	—	4.64
Total	18.875	7	—	—	100.00

+ at least 90% confidence
++ at least 95% confidence
# at least 99% confidence

5-16-1   Water pump experiment

The unpooled and pooled ANOVA summary tables for the water pump experiment are shown in Tables 5-40 and 5-41, respectively. According to the analysis of variance, there are two strong factors which influence water pump leaks. The gasket design and pump finish are the important factors, which agrees with both the observation method and column effects method. ANOVA does not indicate which levels are the best, but indicates that the factors do have a different average result from level to level.

5-16-2   Die-cast piston experiment

The unpooled and pooled ANOVA summary table for the die-cast piston experiment is shown in Tables 5-42 and 5-43, respectively. The ANOVA calculations indicate that there are four factors which influence die-cast piston hardness. Two of the factors are very influential and the other two are relatively weak. The ANOVA calculations do not indicate which levels of the factors are best, only that, statistically, a difference exists between the average results of the levels. The

TABLE 5-42   Die-Cast Piston ANOVA Summary (Unpooled)

Source	SS	v	V	F	P
A	82.688	1	82.688	9.01	5.40
B	58.500	1	58.500	6.37	3.62
$A \times B / D \times E$	58.500	1	58.500	6.37	3.62
C	2.500	1	2.500	0.27	0.00
D	295.000	1	295.000	32.13	21.00
E	487.688	1	487.688	53.12	35.15
$A \times E / B \times D$	9.188	1	9.188	1.00	0.00
e_{reps}	367.250	40	9.181	—	31.21
T	1361.314	47	—	—	100.00

TABLE 5-43   Die-Cast Piston ANOVA Summary (Pooled)

Source	SS	v	V	F	P
A	82.688	1	82.688	9.16#	5.41
B	58.500	1	58.500	6.48++	3.63
$A \times B / D \times E$	58.500	1	58.500	6.48++	3.63
D	295.000	1	295.000	32.70#	21.01
E	487.688	1	487.688	54.05#	35.16
e_p	378.938	42	9.022	—	31.15
T	1361.314	47	—	—	100.00

+ at least 90% confidence
++ at least 95% confidence
# at least 99% confidence

interaction is probably the interaction between die cooling and air cooling, since they are the strongest factors, and not the interaction of copper and magnesium, which are relatively weak factors.

5-16-3   Overall comments on example experiments

The follow-up to these interpretations of experimental results would be to perform a confirmation experiment to verify these conclusions. A confirmation experiment for the water pump example would be to build several assemblies with the new gasket design and smooth pump finish to prove that they would all be leak free.

The confirmation test for the die-cast piston example would be to make several pistons with the air cooling on, the die cooling off, high copper content, and low magnesium content to verify high consistent hardness.

TABLE 5-44   Four-Level Experiment Analysis

		Factors			
	A	B	C	D	E
			Column no.		
Trial no.	1	2	3	4	5	y data
1	1	1	1	1	1	2
2	1	2	2	2	2	6
3	2	1	1	2	2	4
4	2	2	2	1	1	7
5	3	1	2	1	2	7
6	3	2	1	2	1	10
7	4	1	2	2	1	8
8	4	2	1	1	2	12

5-17   Multiple-Level Experiments

This section addresses the analysis of the experimental designs using multiple-level factors, as discussed in Sec. 3-5.

5-17-1   Four-level experiments

The general sums of squares formulas for multiple-level factors used in Sec. 5-8-3 apply in this instance. A four-level factor requires the use of the general formula. Following is a demonstration of the method. The property of orthogonality is undisturbed. If factors A (four-level), B, C, D, and E are assigned to an L8 OA and responses are as shown in Table 5-44, then the analysis for the factors (columns) and total variation is

$$SS_A = \frac{A_1^2}{n_{A_1}} + \frac{A_2^2}{n_{A_2}} + \frac{A_3^2}{n_{A_3}} + \frac{A_4^2}{n_{A_4}} - \frac{T^2}{N}$$

$$= \frac{8^2}{2} + \frac{11^2}{2} + \frac{17^2}{2} + \frac{20^2}{2} - \frac{56^2}{8}$$

$$SS_A = 45.0$$

$$SS_B = \frac{(B_1 - B_2)^2}{N} = \frac{(21 - 35)^2}{8} = 24.5$$

$$SS_C = \frac{(28 - 28)^2}{8} = 0.0$$

$$SS_D = \frac{(28 - 28)^2}{8} = 0.0$$

$$SS_E = \frac{(27-29)^2}{8} = 0.5$$

$$SS_T = 2^2 + 6^2 + 4^2 + 7^2 + 7^2 + 10^2 + 8^2 + 12^2 - \left(\frac{56^2}{8}\right) = 70.0$$

Note that the sums of squares for all the columns add up to the total sum of squares, so the orthogonality of the experiment is still present.

Another demonstration of the preservation of the orthogonality, but not used in a typical experimental analysis, is the fact that the sums of squares for the original three columns that were merged, when added together, equal the sum of squares for factor A.

$$SS_A = SS_{col\ 1} + SS_{col\ 2} + SS_{col\ 3}$$

$$SS_{col\ 1} = \frac{(1_1-1_2)^2}{N} = \frac{(19-37)^2}{8} = 40.50$$

$$SS_{col\ 2} = \frac{(2_1-2_2)^2}{N} = \frac{(25-31)^2}{8} = 4.50$$

$$SS_{col\ 3} = \frac{(3_1-3_2)^2}{N} = \frac{(28-28)^2}{8} = 0.00$$

$$SS_A = 45.00 = 40.50 + 4.50 + 0.00$$

The ANOVA table for the example appears in Table 5-45. A problem is created when interpreting the F ratios relative to each other when two factors have an unequal number of levels. The F ratios seem to indicate that the factor B effect is much larger than the factor A

TABLE 5-45   ANOVA for OA Modified to Four Levels

Source	SS	v	V	F	P
A	45.0	3	15.0	90#	63.6
B	24.5	1	24.5	147#	34.8
C*	0.0	1	0.0		
D*	0.0	1	0.0		
E*	0.5	1	0.5		
T	70.0	7			100.0
e pooled	0.5	3	0.167		1.6

*Indicates those items that were combined to generate the pooled error estimate.
  + at least 90% confidence
  ++ at least 95% confidence
  # at least 99% confidence

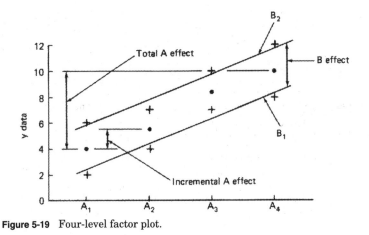

Figure 5-19 Four-level factor plot.

effect, but what does a plot of the data indicate? Figure 5-19 shows the four-level factor data plotted. The F ratios reflect the relative magnitudes of the incremental effects of both factors A and B. Factor A has three increments adding up to the total factor A effect, and factor B has one increment. However, since the total span of level A_1 to A_4 is available from the product or process viewpoint, the total effect of factor A is available also. The total effect is recognized in the percent contribution P for factor A relative to factor B. The percent contribution shows that factor A is almost twice as powerful as factor B for the range of levels used in the experiment.

Care must be exercised when interpreting F values when multiple-level factors and two-level factors are mixed in an experiment. Don't mistake the magnitude of the F ratios to indicate the magnitude of the overall effect. Statistically, the F ratio indicates the presence of a factorial effect, not necessarily the size of the effect. The percent contribution is a much better indicator of relative effects. Also, plots of statistically significant factors will indicate the magnitudes of the effects by the slopes; the higher the slope, the greater the factor effect. The F ratio basically indicates which factors and interactions should be plotted.

Other interpretations that may be made of the plot depend upon the type of response; lower is better (LB), nominal is best (NB), or higher is better (HB). If an HB characteristic, the treatment condition of A_4B_2 is the best within the investigated experimental space. If an LB characteristic, the treatment condition of A_1B_1 is the best within the experimental space. Either one of these characteristics would suggest further experimentation to increase or decrease the average result for an HB or LB characteristic, respectively. An NB character-

TABLE 5-46   Polynomial Effects for
Increasing Levels

No. levels	Polynomial effect
2	linear
3	quadratic
4	cubic

istic could be interpolated fairly accurately to estimate a particular
treatment condition that would provide the proper average.

5-17-2   Polynomial decomposition

When specific conditions are met in an experiment, the factor sum of
squares may be decomposed into even smaller sources. If the factor
evaluated is of a continuous nature having equal increments between
levels and an equal sample size at all levels, then polynomial decomposition is possible using a simple method. Table 5-46 shows the various polynomial effects that may be estimated. The linear and quadratic effects are frequently found, although a factor with a pure
cubic effect is rarely seen.

Another example will be used to demonstrate the polynomial
decomposition method. Table 5-47 shows some possible results for an
L8 OA modified to a four-level first-column arrangement (remaining

TABLE 5-47   Four-Level Experimental Data

Trial no.	A	y data
1	1	4
2	1	5
3	2	8
4	2	9
5	3	10
6	3	9
7	4	5
8	4	4

TABLE 5-48   Components of Four-
Level Factor Variation

Source	v
$SS_{A\ linear}$	1
$SS_{A\ quadratic}$	1
$SS_{A\ cubic}$	1
$SS_{A\ total}$	3

columns are omitted for clarity). The sum of squares for factor A is

$$SS_A = \frac{A_1^2}{n_{A_1}} + \frac{A_2^2}{n_{A_2}} + \frac{A_3^2}{n_{A_3}} + \frac{A_4^2}{n_{A_4}} - \frac{T^2}{N}$$

$$SS_A = \frac{9^2}{2} + \frac{17^2}{2} + \frac{19^2}{2} + \frac{9^2}{2} - \frac{54^2}{8}$$

$$SS_A = 41.5$$

The SS_A value has 3 degrees of freedom associated with it. Those 3 degrees of freedom can be broken down into the components shown in Table 5-48. Each of the degrees of freedom may be used to estimate a polynomial effect. The number of polynomial effects that may be estimated is equal to $k-1$ (where k equals the number of levels), the degrees of freedom for that factor. A two-level factor may only have the linear effect estimated, a three-level factor may have the linear and quadratic effects estimated, etc.

The method for calculating the sum of squares for the various polynomial effects is based on the same equation.

$$SS_{A\,polynomial\,effect} = \frac{(W_1 A_1 + W_2 A_2 + \dots + W_k A_k)^2}{W_T R}$$

Table D-5 in App. D contains the W coefficients that are appropriate for a particular polynomial effect. The table is set up for two- to five-level factors. The R value is equal to the number of samples under a level. In this example, the W coefficients are taken from the four-level column. The various polynomial effects are then calculated:

$$SS_{A\,linear} = \frac{[-3(9) - 1(17) + 1(19) + 3(9)]^2}{20(2)} = 0.1$$

$$SS_{A\,quadratic} = \frac{[1(9) - 1(17) - 1(19) + 1(9)]^2}{4(2)} = 40.5$$

$$SS_{A\,cubic} = \frac{[-1(9) + 3(17) - 3(19) + 1(9)]^2}{20(2)} = 0.9$$

$$SS_A = SS_{A\,linear} + SS_{A\,quadratic} + SS_{A\,cubic} = 41.5$$

This example demonstrates how the sums of squares can be decomposed into polynomial effects. Also, the different polynomial effects can be F-tested to determine if any are statistically significant. In this example, the quadratic effect is much larger than any other. Figure 5-20 shows the relative magnitudes. The data is sloped slight-

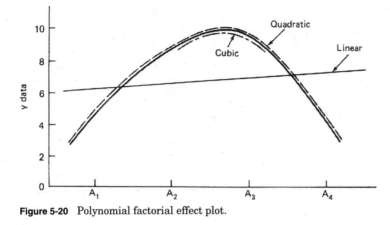

Figure 5-20   Polynomial factorial effect plot.

ly up to the right, which is proportional to the linear effect of factor A. The graph has one maximum and is greatly curved, which is proportional to the quadratic effect of factor A. And the slight rate of change of the curvature is proportional to the cubic effect of factor A. If the A_3 data points were values of 8 and 9, there would be no linear or cubic effect but purely a quadratic effect present. A parabola (quadratic equation) would fit the averages of the data at the four levels exactly. The two data points at each level of factor A imply the error variation that is present around each of the level averages. This is one method to determine if the data, such as in Sec. 5-8-8 and plotted in Fig. 5-16, follows a curved or straight line. The smaller insignificant polynomial effects can be pooled for error estimation with other factors if necessary.

Remember, this polynomial decomposition method depends on

1. Continuous variables
2. Equal increments between levels
3. Equal sample sizes at all levels

5-17-3   Three-level factors (dummy treatment)

Recalling the approach of adapting a four-level column to handle a three-level factor in Sec. 3-5-4, the degrees of freedom indicate that there is more information available in the column than just factor

TABLE 5-49   Experimental Results Using Dummy Treatment

	Factors					
	A	B	C	D	E	
	Column no.					
Trial no.	1	2	3	4	5	y data
1	1	1	1	1	1	10
2	1	2	2	2	2	5
3	2	1	1	2	2	8
4	2	2	2	1	1	4
5	3	1	2	1	2	1
6	3	2	1	2	1	7
7	$1'$	1	2	2	1	6
8	$1'$	2	1	1	2	11

effects. The column to which factor A, a three-level factor, is assigned has 3 degrees of freedom associated with it. Factor A, however, has only 2 degrees of freedom, so the remaining unassigned degree of freedom is associated with error. Because of this dual source of variation, the ANOVA must be performed in a slightly different manner. An example will facilitate understanding of the ANOVA method. Table 5-49 shows some possible experimental results. The formula for sum of squares due to factor A must use the general form

$$\mathrm{SS}_A = \frac{A_1^2}{n_{A_1}} + \frac{A_2^2}{n_{A_2}} + \frac{A_3^2}{n_{A_3}} - \frac{T^2}{N}$$

The level symbols for factor A_1 and $A_{1'}$, both indicate the same test condition with respect to factor A. Therefore,

$$\mathrm{SS}_A = \frac{(A_1 + A_{1'})^2}{n_{A_1} + n_{A_{1'}}} + \frac{A_2^2}{n_{A_2}} + \frac{A_3^2}{n_{A_3}} - \frac{T^2}{N}$$

$$= \frac{(15 + 17)^2}{2 + 2} + \frac{12^2}{2} + \frac{8^2}{2} - \frac{52^2}{8} = 22.00$$

The variation due to error can be treated in the same manner as a two-level factor, that is, a comparison of the averages under the 1 and $1'$ conditions. The denominator of the fraction is still the total number of tests involved in that comparison.

Where $n_{A_1} = n_{A_{1'}}$

$$\mathrm{SS}_e = \frac{(A_1 - A_{1'})^2}{n_{A_1} + n_{A_{1'}}}$$

TABLE 5-50   ANOVA Summary for Dummy Treatment

Source	SS	v	V	F	P
A	22.00	2	11.00	22.0#	28.4
B*	0.50	1	0.50		
C	50.00	1	50.00	100.0#	66.9
D*	0.00	1	0.00		
E*	0.50	1	0.50		
e*	1.00	1	1.00		
T	74.00	7			100.0
e pooled	2.00	4	0.50		4.7

*Indicates those items that were combined to generate the pooled error estimate.
+ at least 90% confidence
++ at least 95% confidence
# at least 99% confidence

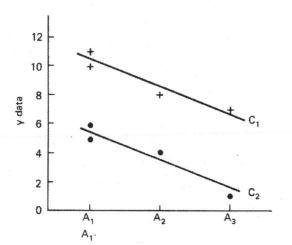

Figure 5-21 Dummy treatment plot.

$$SS_e = \frac{(15 - 17)^2}{4} = 1.00$$

As an exercise, the remaining sources of variation and total variation should be calculated for the example. The total variation can still be accounted for in the columns. Table 5-50 summarizes the ANOVA calculations for this example.

A plot of the significant factors in Fig. 5-21 shows how the dummy treatment is used and causes an unequal number of samples in the different test conditions. The orthogonality of the experiment is still

TABLE 5-51   Nested Factor Experiment Data

			Factors				
	A	B,E		C,e	D	A × D	Force,
			Column no.				lb
Trial no.	1	2		3	4	5	
1	1	1		1	1	1	4
2	1	1	B	2	2	2	10
3	1	2		1	2	2	14
4	1	2		2	1	1	8
5	2	1		1	1	2	3
6	2	1	E	2	2	1	8
7	2	2		1	2	1	7
8	2	2		2	1	2	2

Note: 1 lb = 0.454 kg; e = error.

intact, however, because the total of the sums of squares for the merged columns now equals the sum of squares for factor A plus the error sum of squares.

$$SS_A + SS_e = SS_{col\,1} + SS_{col\,2} + SS_{col\,3}$$

5-18   Special Designs

This section covers the analysis of the special experimental designs discussed in Sec. 3-6, which were nested factors, combined factors, and the idle column method.

5-18-1   Nested factors

Recalling the nesting/nested factor discussion in Sec. 3-6-1, an error estimate will come from the half of the experiment when the parts are press fit, level 2 for factor A, and the levels indicating the temperature of heat staking, factor C, are meaningless. Table 3-18 shows this assignment and location of the error estimate.

If the press-out force were measured for each of the assemblies and had values as in Table 5-51, the analysis for the nesting, common factors, and interactions would be

$$SS_A = \frac{(A_1 - A_2)^2}{N} = \frac{(36 - 20)^2}{8} = 32.00$$

$$SS_D = \frac{(D_1 - D_2)^2}{N} = \frac{(17 - 39)^2}{8} = 60.50$$

$$SS_{A \times D} = \frac{(A \times D_1 - A \times D_2)^2}{N} = \frac{(27 - 29)^2}{8} = 0.50$$

The interactions between nesting and nested factors are not included in the OA or analysis because the interactions are nonexistent. When a certain level of the nesting factor is chosen, there are only two conditions for the nested factor and these conditions do not exist for the other level of the nesting factor.

The analysis for the nested factors must accommodate the smaller sample size within the nest and the error estimate in a portion of a column:

$$SS_B = \frac{(B_1 - B_2)^2}{(n_{B_1} + n_{B_2})} = \frac{(14 - 22)^2}{(2 + 2)} = 16.00$$

$$SS_C = \frac{(C_1 - C_2)^2}{(n_{C_1} + n_{C_2})} = \frac{(18 - 18)^2}{(2 + 2)} = 0.00$$

$$SS_E = \frac{(E_1 - E_2)^2}{(n_{E_1} + n_{E_2})} = \frac{(11 - 9)^2}{(2 + 2)} = 1.00$$

$$SS_{e \, col \, 3} = \frac{(3_1 - 3_2)^2}{(n_{3_1} + n_{3_2})} = \frac{(10 - 10)^2}{(2 + 2)}$$

$$SS_{e \, col \, 3} = 0.00$$

Table 5-52 shows the ANOVA summary for this particular set of experimental conditions and data. Figure 5-22 shows individual plots for the three significant factors. Since press-out force is an HB char-

TABLE 5-52   ANOVA Summary for a Nested Experiment

Source	SS	v	V	F
A	32.0	1	32.0	84.21#
B	16.0	1	16.0	42.11#
C*	0.0	1	0.0	
D	60.5	1	60.5	159.21#
E*	1.0	1	1.0	
A×D*	0.5	1	0.25	
T	110.0	7		
e	0.0	1	0.0	
e pooled	1.5	4	0.38	

*Indicates those items that were combined to generate the pooled error estimate.
 + at least 90% confidence
 ++ at least 95% confidence
 # at least 99% confidence

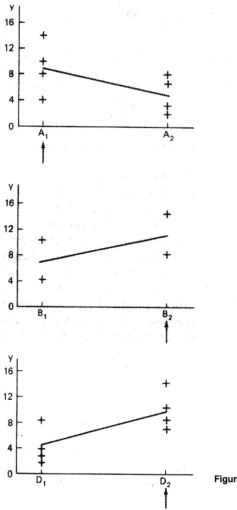

Figure 5-22 Nested factors plots.

acteristic, the treatment condition of $A_1B_2D_2$ is thought to provide the maximum value of the force.

Taguchi uses the pooling-up strategy for F-testing the remaining factors in the experiment. Traditionally, however, statisticians recommend using the error variance within a nest (group) to F-test the factors within the nesting factor. It is very possible that the two retention methods do not have equal variance, which violates one of the assumptions of ANOVA. Taguchi is trying to identify from a practical viewpoint which factors affect performance and uses the F test as a reference value. The confirmation experiment is used to validate the combined effects of the factors selected.

TABLE 5-53   Combined Factors Experiment Data

	Factors				
	AB	C	D	E	
	Column no.				
Trial no.	1	2	3	4	Data
1	1	1	1	1	2
2	1	2	2	2	6
3	1	3	3	3	11
4	2	1	2	3	6
5	2	2	3	1	10
6	2	3	1	2	1
7	3	1	3	2	14
8	3	2	1	3	6
9	3	3	2	1	10

5-18-2   Combined factors

The main limitations of the combined factors approach is the loss of the ability to estimate the interaction that may exist between the two combined factors and the loss of orthogonality of the experiment. This approach was introduced in Sec. 3-6-2.

An example L9 OA and data demonstrate the impact of the lost orthogonality. Table 5-53 shows an L9 combined factor column assignment and test results. The sum of squares calculations for the total variation and variation due to factors are

$$SS_T = 2^2 + 6^2 + \ldots + 10^2 - \left(\frac{66^2}{9}\right) = 146.00$$

$$SS_{AB} = \frac{(AB)_1^2}{n(AB)_1} + \frac{(AB)_2^2}{n(AB)_2} + \frac{(AB)_3^2}{n(AB)_3} - \frac{T^2}{N}$$

$$SS_{AB} = \frac{19^2}{3} + \frac{17^2}{3} + \frac{30^2}{3} - \frac{66^2}{9} = 32.67$$

$$SS_C = \frac{22^2}{3} + \frac{22^2}{3} + \frac{22^2}{3} - \frac{66^2}{9} = 0.0$$

$$SS_D = \frac{9^2}{3} + \frac{22^2}{3} + \frac{35^2}{3} - \frac{66^2}{9} = 112.67$$

$$SS_E = \frac{22^2}{3} + \frac{21^2}{3} + \frac{23^2}{3} - \frac{66^2}{9} = 0.66$$

The total orthogonality has not been destroyed:

$$SS_T = SS_{AB} + SS_C + SS_D + SS_E$$
$$146.00 = 32.67 + 0.0 + 112.67 + 0.67$$

The sum of squares for factor A can be calculated for the comparison of $(AB)_1$ to $(AB)_2$ by applying the specific two-level formula

$$SS_A = \frac{[(AB)_1 - (AB)_2]^2}{n_{(AB)_1} + n_{(AB)_2}}$$

$$= \frac{[19 - 17]^2}{3 + 3} = 0.67$$

Similarly, for factor B,

$$SS_B = \frac{[(AB)_2 - (AB)_3]^2}{n_{(AB)_2} + n_{(AB)_3}}$$

$$= \frac{[17 - 30]^2}{3 + 3} = 28.17$$

As a demonstration of the lost orthogonality between factors A and B, the following inequality can be written.

$$SS_{AB} \neq SS_A + SS_B$$
$$32.67 \neq 0.67 + 28.17$$

Therefore, the orthogonality between factors A and B is lost.

In this situation, the factor A effect is very small and the factor B effect is very large; hence, factor B must be the significant factor of the two. If this is true, then the combined factors treatment is equivalent to a dummy treatment of factor B. With a dummy treatment, the orthogonality of the entire experiment is preserved. Level $(AB)_2$ can be thought of as the dummy level for factor B which would make that level the $B_{1'}$. Calculating the sum of squares for factor B as if it were dummy-treated, which assumes the factor A effect is zero, would give these results:

$$SS_B = \frac{(B_1 + B_{1'})^2}{n_{B_1} + n_{B_{1'}}} + \frac{B_2^2}{n_{B_2}} - \frac{T^2}{N}$$

$$= \frac{(19 + 17)^2}{3 + 3} + \frac{30^2}{3} - \frac{66^2}{9} = 32.00$$

$$SS_e = \frac{(B_1 - B_{1'})^2}{n_{B_1} + n_{B_{1'}}}$$

$$SS_e = \frac{(19 - 17)^2}{(3 + 3)} = 0.67$$

So, therefore,

$$SS_{AB} = SS_B + SS_e$$

$$32.67 = 32.00 + 0.67$$

Since the entire experiment is orthogonal for a dummy treatment,

$$SS_T = SS_A + SS_B + SS_C + SS_D + SS_E + SS_e$$
$$SS_A = 0.0$$
$$146.00 = 0.0 + 32.00 + 0.0 + 112.67 + 0.66 + 0.67$$

This can be done only if the factor A or B effect is very small relative to the other combined factor effect.

5-18-3   Idle column method

This approach was introduced in Sec. 3-6-3 when three-level factors were used in a two-level OA. The analysis of a factor using an idle column assignment is done in two parts. $SS_{A(1-2)}$ and $SS_{A(2-3)}$ are added together to obtain the total SS_A effect. However, because of using the idle column method, part of factor A's effect is in the idle column and is not able to be estimated. When several factors are associated with the idle column, a mixture of parts of each of the factorial effects plus error effects are in the idle column. Because of this confounding of factorial and error effects, a factor should not be assigned to the idle column unless it is a nuisance variable that is to be blocked in the experiment.

TABLE 5-54   Idle Column Experiment Example

		Factors				
	Idle	A	B	C	D	
			Column no.			
Trial no.	1	2	3	4	5	Data
1	1	1	1	1	1	3
2	1	1	2	2	2	6
3	1	2	1	2	2	13
4	1	2	2	1	1	6
5	2	2	2	1	2	8
6	2	2	3	2	1	11
7	2	3	2	2	1	12
8	2	3	3	1	2	9

TABLE 5-55   ANOVA Summary for Idle Column Experiment

Source	SS	v	V	F
Idle	18.0	1	18.0	10.3++
$A_{(1-2)}$	25.0	1	25.0	14.3++
$A_{(2-3)}$*	1.0	1	1.0	
$B_{(1-2)}$*	4.0	1	4.0	
$B_{(2-3)}$*	0.0	1	0.0	
C	32.0	1	18.3++	
D*	2.0	1	2.0	
T	82.0	7		
e pooled	7.0	4	1.75	

*Indicates those items that were combined to generate the pooled error estimate.
+ at least 90% confidence
++ at least 95% confidence
# at least 99% confidence

Using an idle column assignment and data as shown in Table 5-54, the sum of squares calculations are

$$SS_T = 82.0$$

$$SS_{A_{(1-2)}} = \frac{(9-19)^2}{4} = 25.0$$

$$SS_{A_{(2-3)}} = \frac{(19-21)^2}{4} = 1.0$$

$$SS_{B_{(1-2)}} = \frac{(16-12)^2}{4} = 4.0$$

$$I_1 \begin{cases} \overline{A}_1 = \dfrac{9}{2} = 4.5 \\[2em] \overline{A}_2 = \dfrac{19}{2} = 9.5 \end{cases}$$

$$\left.\begin{matrix} \\ \\ \\ \\ \end{matrix}\right\} \overline{A}_2 = \frac{38}{4} = 9.5$$

$$I_2 \begin{cases} \overline{A}_2 = \dfrac{19}{2} = 9.5 \\[2em] \overline{A}_3 = \dfrac{21}{2} = 10.5 \end{cases}$$

$$SS_{B_{(2-3)}} = \frac{(20-20)^2}{4} = 0.0$$

$$SS_C = \frac{(26-42)^2}{8} = 32.0$$

$$SS_D = \frac{(32-36)^2}{8} = 2.0$$

$$SS_{idle} = \frac{(28-40)^2}{8} = 18.0$$

The ANOVA summary is shown in Table 5-55.

Here, factor $A_{(1-2)}$ is significant but $A_{(2-3)}$ is not, which indicates some difference in averages from condition A_1 to A_2, but no detectable difference in averages from A_2 to A_3.

The comparison of levels A_1 to A_2 involved a total of four samples, as did the comparison of A_2 to A_3 when the idle column indicated levels 1 and 2, respectively.

A plot (not shown) of the three levels of factor with two comparisons would indicate this possibly to be a nonlinear response situation. If this were an HB characteristic, then level A_3 would have the highest average and could be used, depending on economics, even though statistically no different than A_2. If the A_3 condition is a cost penalty, then perhaps A_2 should be used to obtain nearly the same performance at a lower cost.

5-19   Attribute Data

Although designing an experiment is the same regardless of the type of data, variable or attribute, the application of ANOVA is slightly dif-

TABLE 5-56   Attribute Data Analysis Methods

	Known occurrences and nonoccurrences	Known occurrences only
Two classes	Case I ANOVA frequency of occurrence % occurrence 0, 1 data	Case III ANOVA frequency of occurrenc
More than two classes	Case II Accumulation analysis	Case IV Not discussed

ferent with attribute data. Attribute data may be analyzed many ways, depending on the type of attribute data (known occurrences and nonoccurrences or known occurrences only) and how many classes are used (two or more than two). The kind of analysis and the kind of data for the situations described are shown in Table 5-56. The most used cases are I and II, case III is used occasionally, and case IV is seldom used. Cases I through III will be discussed in Secs. 5-19-1 through 5-19-3, respectively, but case IV will not be discussed.

5-19-1   Case I: casting cracks with two classes

Two-class attribute data is again the situation corresponding to an all-or-nothing characteristic, which can be represented numerically by a 1 or 0. In this case, the number of occurrences in both classes is known. The primary analytical approach should use the number of occurrences of defective parts treated as a lower-is-better characteristic. This case study contains some testing philosophy as well as the Taguchi analytical approach.

Casting problem background.   An actual production problem which required the use of attribute data was one concerning cracks in a main structural casting for a heavy-duty automatic transmission. The cracks had not always been a problem but had developed into a 10 to 15% defective castings problem in recent times. Something had definitely changed in the process, but try as they might, the engineers could not resolve the cracking problem. Cracks were sometimes visible with the naked eye, but most were only discernible with magnaflux inspection. Cracks were also more easily noticed after the part had been machined, which made the scrap expense even greater. Recall, from the loss function discussion, that the sooner a problem is discovered, the lower the loss associated with it. Cracks were being found in various locations around the perimeter of the wagon wheel–shaped part; no pattern of the location was apparent.

One hundred percent magnaflux inspection was instituted as a stopgap measure. The expensive magnaflux process was slow, and the scrap rate of castings began to interrupt manufacturing and assembly lines. The economics of magnaflux inspection, $2 per $17 casting, plus the schedule impact, made resolution of this problem very desirable.

Casting cracks experiment structure.   A meeting was held to discuss the possible use of a designed experiment to find a permanent solution. The meeting included two casting experts from the manufacturing plant, a foundry representative, a product design engineer, a purchasing agent, and an engineer trained in statistical methods. The focus of the discussion was on what factors might be causing the

TABLE 5-57   Casting Cracks Experiment Factors and Levels

Factor	Level 1	Level 2
A Temperature	Production	Higher
B Time	Production	Shorter
C Cooling rate	Production (fan on)	Fan off
D Shot blast	Production (one cycle)	More cycles

cracks and the structure of an experiment for investigating these factors. The foundry people were convinced that the use of Taguchi OAs was too complex for the foundry environment and wanted to run simpler two-factor experiments. Other roadblocks were thrown up to make their case.

The factors of interest were pouring temperature, time in the mold, cooling rate after removal from the mold, and shot blast intensity. To investigate all these factors, two at a time, for all possible combinations required six different experiments with 24 different treatment conditions. It was pointed out that an L8 OA could investigate all these things with only eight different treatment conditions. The casting engineers asked how each casting in a trial could be poured at a particular temperature when temperature would fall as castings were poured from the ladle. Data was available which indicated a 40°F temperature drop from the first casting poured to the 12th and final casting poured. The high and low pour temperatures were revised to indicate the temperature at the start of pouring when the ladle was removed from the holding furnace. If one temperature was better than the other, then this would be the way the foundry would process the castings anyway; one casting at a time would be highly impractical. This procedure introduced some setting error into the experiment but, again, the foundry had to live with this on a day-to-day basis anyway. These were the major roadblocks that were resolved, and agreement was made to go ahead with the experiment.

The next step was to select the levels for the casting factors. A summary of the factors and the levels suggested is in Table 5-57. The actual values for the temperature, time, etc., were determined but not recorded here for proprietary reasons.

The number of castings needed per trial can be determined by dividing the minimum of 20 defectives for the whole experiment by the past defective rate; 20 divided by 0.10 portion defective gives a minimum of 200 castings required for the whole experiment. Since 12 castings are obtained per ladle, then 16 ladles would make 192 castings. The 16 ladles would conveniently adapt to an L8 or L16 OA.

Foundry preexperiment.   As a historical note, the casting experts in conjunction with the foundry felt that the cracks were caused by pouring the castings too hot, knocking them out of the mold too soon,

TABLE 5-58   Casting Cracks OA Assignment

							No. of castings	
				Factors and interactions				
	A	B	$A \times B$ $C \times D$	C	$A \times C$ $B \times D$	$A \times D$ $B \times C$	D		
				Column no.				No. of castings	
Trial no.	1	2	3	4	5	6	7	Cracked	Good
1	1	1	1	1	1	1	1	4	20
2	1	1	1	2	2	2	2	1	23
3	1	2	2	1	1	2	2	1	23
4	1	2	2	2	2	1	1	2	22
5	2	1	2	1	2	1	2	4	20
6	2	1	2	2	1	2	1	4	20
7	2	2	1	1	2	2	1	0	24
8	2	2	1	2	1	1	2	0	24
							Totals	16	176

and cooling them too quickly. To prove their conjecture was correct before the entire experimental effort was wasted, the foundry poured 20 castings at a high temperature, knocked them out of the mold as soon as possible, and threw them into a snow drift to accelerate cooling. All 20 of the castings were good, being entirely crack free. The experts were quoted as saying "Well, they all didn't crack in production, you know." This was a clue about the casting process, but with no other conditions to reference, the understanding is consequently minimal.

Assignment to an orthogonal array. The factors were assigned to an L8 OA with the assignment and experimental results shown in Table 5-58. Recall that two ladles, 24 castings, were poured for each of the trial conditions.

Observation method analysis. In this experiment, trials 7 and 8 provided good results relative to the rest of the experiment and recent production. Note, the common levels for temperature, time, and, consequently, the interaction of temperature and time. The high pouring temperature in conjunction with short mold time appear to provide good results consistent with the "snow drift" experiment and contrary to prior opinions.

Analysis using frequency of occurrence. The most meaningful ANOVA that may be performed is one using the frequency of occurrence of cracked castings within a trial. This approach will be identical analytically with variable data. The magnitude of frequency of occurrence is a measure of how poor a given process condition is in providing good castings. The data takes the form of integers, since a casting is either good or bad. The ANOVA summary is shown in Table

TABLE 5-59   ANOVA Summary Using Frequency of Occurrence

Source	SS	v	V	F	P
A^*	0.00	1	0.00		
B	12.50	1	12.50	12.50++	52.3
C^*	0.50	1	0.50		
D^*	2.00	1	2.00		
$A{\times}B/C{\times}D$	4.50	1	4.50	4.50+	15.9
$A{\times}C/B{\times}D^*$	0.50	1	0.50		
$A{\times}D/B{\times}C^*$	2.00	1	2.00		
T	14.667	191			100.00
e pooled	5.00	5	1.00		31.8

*Indicates those items that were combined to generate the pooled error estimate.
+ at least 90% confidence
++ at least 95% confidence
# at least 99% confidence

5-59. For edification, the ANOVA table would be identical if the count of occurrences of good castings had to be analyzed rather than the count of occurrences of bad castings.

This analysis shows that factor B, mold time, has the single biggest effect and contributes a high percentage to the total observed variation. The interaction column containing $A{\times}B$ or $C{\times}D$ has the next biggest effect and contributes a substantial amount to the total variation. In this situation, the interaction is more likely to be the $A{\times}B$ interaction since factor B is the strongest single factor in the experiment and the two best trials had common levels of temperature, time, and the temperature/time interaction (observation method).

The percent contribution should be carefully interpreted in attribute data analysis. There is a substantial amount of variation associated with error, which implies that there will still be some variation under the best conditions. Recall that the error variation is calculated by pooling all the error sums of squares together and dividing by the pooled degrees of freedom. This is the foundation for the ANOVA assumption of equal variance in all the treatment conditions. With attribute data, however, there may be a condition which provides all good products with no variation. The error variation comes from other treatment conditions that provide some good and some bad products, which is the attribute form of variation. A confirmation test is needed to verify that a certain test condition is defect free.

Analysis using 0, 1 data.   Another form of ANOVA for two-class attribute data is as follows. Recall that the bad and good parts can be mathematically represented by 1s and 0s, respectively. Therefore, the data for the first trial is really four 1s and twenty 0s. If 1s and 0s are considered as the data, then the total sum of squares is

$$SS_T = 1^2 + 1^2 + 1^2 + 1^2 + 0^2 + 0^2 + \ldots + 0^2 - \left(\frac{16^2}{192}\right)$$

Simplified, because T = sum of all data, $1^2 = 1$, and $0^2 = 0$

$$SS_T = T - \left(\frac{T^2}{N}\right) = 16 - \left(\frac{16^2}{192}\right) = 14.667$$

The sums of squares for the columns are calculated as with variable data:

$$SS_A = \frac{(8 - 8)^2}{192} = 0.0$$

$$SS_B = \frac{(13 - 3)^2}{192} = 0.521$$

$$SS_C = \frac{(9 - 7)^2}{192} = 0.021$$

$$SS_D = \frac{(10 - 6)^2}{192} = 0.083$$

$$SS_{A \times B / C \times D} = \frac{(5 - 11)^2}{192} = 0.188$$

$$SS_{A \times C / B \times D} = \frac{(9 - 7)^2}{192} = 0.021$$

$$SS_{A \times D / B \times C} = \frac{(10 - 6)^2}{192} = 0.083$$

There are two sources of error in this experiment which need to be identified. The first source of error will be estimated from the columns (factors and interactions are assigned but have negligible effects) of the OA. The second source, labeled e, is from the repetitions in each trial. Intuitively, a form of error can be identified in trial 1, for example. All 24 castings were processed under very similar conditions, yet not all are good or all cracked. If no error were present, then the 24 castings in each trial would be either all good or all bad. Mathematically, the SS_e amount can be calculated by finding the average of the 0, 1 data for each trial, squaring the deviation of the 0, 1 data from the average, and summing the squares of the deviations. This is nothing more than an extension of the ANOVA method learned earlier in this chapter (error deviations within groups). In this situation, the total variation, SS_T, can be calculated and the variation due to each accountable source, SS_{cols}, subtracted to obtain the variation

TABLE 5-60   ANOVA Summary Using 0,1 Data

Source	SS .	v	V	F	P
A^*	0.0	1	0.0		
B	0.521	1	0.521	6.95#	3.05
C^*	0.021	1	0.021		
D^*	0.083	1	0.083		
$A{\times}B/C{\times}D$	0.188	1	0.188	2.51	0.78
$A{\times}C/B{\times}D^*$	0.021	1	0.021		
$A{\times}D/B{\times}C^*$	0.083	1	0.083		
e^*	13.750	184	0.075		
T	14.667	191			100.00
e pooled	13.958	189	0.074		96.17

*Indicates those items that were combined to generate the pooled error estimate.
+ at least 90% confidence
++ at least 95% confidence
# at least 99% confidence

leftover for repetition error, SS_e. The degrees of freedom associated with error are determined in a similar manner—the accountable degrees of freedom are subtracted from the total available:

$$v_e = v_T - v_{\text{columns}} = 191 - 7 = 184$$

A summary of the ANOVA results is shown in Table 5-60. From a practical point of view, the error variance usually doesn't need to be pooled with this type of data analysis. The degrees of freedom associated with error are usually very large and the error variance is changed

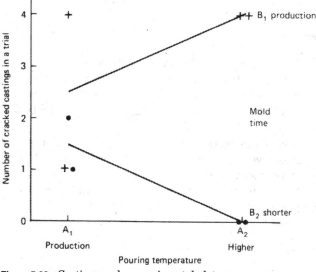

Figure 5-23   Casting cracks experimental plot.

very little when pooled. Also, the percent contribution may calculate to be a small value, which can be misleading. In this case, if the proper temperature and time combination provide consistently good results, the error under other factor combinations is irrelevant, so care must be taken when interpreting the percent contribution, which is a function of the average factorial effect minus the averaged error effect.

This interpretation using approach to ANOVA is consistent with the previous approach, with the exception of the percent contribution. Time by itself causes a change in defective rate at a high statistical confidence, and temperature by itself causes no change in defective rate. The temperature-time interaction is not quite at the 90% confidence level. A plot of the interaction will clarify the interpretation. Note that only one of the factorial combinations is consistent; A_2B_2 has all good parts, while the others are some portion good and bad as shown in Fig. 5-23. The error variation averaged over all the treatment conditions masks the contribution of the temperature-time combination. But from a practical point of view, temperature and time actually provide a tremendous contribution toward reducing defective castings when the proper levels are used. This is particularly reinforced by the observation method mentioned earlier.

Casting experiment summary. In retrospect, the snow drift experiment that had been run earlier is understandable. The conditions in the snow drift experiment for pour temperature and mold time were very similar to conditions in trials 7 and 8 for the L8 OA. The OA experiment, however, allowed the relationship between temperature and time to be comprehended. The casting experts were so intent on a preconceived notion of which factors and conditions were causing a problem that they were preventing themselves from deriving a solution. In fact, the production conditions for temperature and time suggested by the engineers who were led by their experience and intuition caused 10.4% (5 out of 48) of the castings to crack in the experiment. This was very similar to the recent production history. One advantage of an OA experiment is its objectivity. Some combinations of factors and levels are tested which otherwise may not have been investigated. Evidently, the foundry had lost control of temperature and/or time, which caused some bad castings to show up. When the casting experts took over and used experience of other casting problems to adjust the process, the defect problem actually became worse.

Another aspect of these experimental results is that the casting experts could not explain why the pouring conditions of higher temperature and shorter time prevented cracks. The knowledge of why cracks are caused or why they can be prevented is irrelevant if what to do to prevent cracks is known. Even the fact that an interaction between temperature and time may exist becomes irrelevant.

Knowing what temperature and time consistently prevents cracks is the practical information that can be put to use on the factory floor. Of course, if the reason why can be determined, then this information can be used in other casting designs and development. This critique of the casting experts is not meant to belittle their knowledge or experience; one must understand the physics of a design or process to know which factors to investigate. Understanding the problem areas will, hopefully, encourage the use of more effective experimentation methods when faced with a problem.

Percent of occurrences. Case I mentions transformation of attribute data to other forms of data which become variable and can be analyzed by the typical ANOVA methods of Chap. 2. The actual frequency of occurrences in a trial is a discrete variable; the percent of occurrences in a trial is a continuous variable within the limits of 0.0 to 100.0%. This approach to analysis will provide results very similar to the frequency of occurrence analysis previously discussed.

5-19-2   Case II: castings cracks with multiple classes

The data for the casting experiment was also collected in a fashion which described the severity of the cracks that appeared in the casting. Accumulation analysis takes frequency and severity of occurrences into account and is used when the class number has an engineering meaning. The classes for crack severity can be described along with the accompanying rating value:

Severity	Class number
No cracks	1
Mild crack(s)	2
Moderate crack(s)	3

TABLE 5-61   Frequency of Occurrence of Castings

Trial no.	Class 1	2	3	4	Totals
1	20	2	1	1	24
2	23	0	0	1	24
3	23	0	1	0	24
4	22	0	2	0	24
5	20	4	0	0	24
6	20	2	2	0	24
7	24	0	0	0	24
8	24	0	0	0	24
Totals	176	8	6	2	192

TABLE 5-62   Cumulative Frequency of Occurrence of Castings

Trial no.	Cumulative class			
	I	II	III	IV
1	20	22	23	24
2	23	23	23	24
3	23	23	24	24
4	22	22	24	24
5	20	24	24	24
6	20	22	24	24
7	24	24	24	24
8	24	24	24	24
Totals	176	184	190	192

Severe crack(s)	4

The frequency of the castings falling into these various categories is then tallied in Table 5-61. Trial 6 resulted in 20 castings with no cracks, 2 castings with mild cracks, and 2 castings with moderate cracks. The other trials are interpreted accordingly.

The first step in accumulation analysis is to create a cumulative frequency table by summing the frequencies left to right (or right to left) for each trial. The cumulative values are calculated:

Class I = class 1

Class II = class 1 + class 2

Class III = class 1 + class 2 + class 3

Class IV = class 1 + class 2 + class 3 + class 4

Table 5-62 shows the cumulative frequencies for the classes which were determined by summing from left to right. One will notice that the last class shows no variation from trial to trial. For this reason, in accumulation analysis, the number of classes minus one is analyzed.

The second step is to calculate a weight value for each class, which is a function of the cumulative frequency of occurrence in that class. The formula for the weight for class i is

$$W_i = \frac{N^2}{T_i(N - T_i)}$$

where N = total number of tests

T_i = accumulative occurrences in class i

Therefore,

$$W_{\mathrm{I}} = \frac{N^2}{T_{\mathrm{I}}(N - T_{\mathrm{I}})} = \frac{192^2}{176(192 - 176)} = 13.09$$

$$W_{\mathrm{II}} = \frac{N^2}{T_{\mathrm{II}}(N - T_{\mathrm{II}})} = \frac{192^2}{184(192 - 184)} = 25.04$$

$$W_{\mathrm{III}} = \frac{N^2}{T_{\mathrm{III}}(N - T_{\mathrm{III}})} = \frac{192^2}{190(192 - 190)} = 97.01$$

The sum of squares for each class is multiplied by the class weight to obtain a weighted sum of squares for each class. The weighted sums of squares are then added to obtain the complete sum of squares.

The third step is to calculate the total sum of squares:

$$\mathrm{SS}_T = \sum_{i=1}^{\mathrm{nca}} \mathrm{SS}_{T_i} W_i$$

where nca is the number of classes analyzed. Similar to two-class analysis,

$$\mathrm{SS}_{T_i} = T_i - \frac{T_i^2}{N}$$

Then in this example,

$$\mathrm{SS}_T = \mathrm{SS}_{T_{\mathrm{I}}} W_{\mathrm{I}} + \mathrm{SS}_{T_{\mathrm{II}}} W_{\mathrm{II}} + \mathrm{SS}_{T_{\mathrm{III}}} W_{\mathrm{III}}$$

Substituting, but showing the detail in the first class only,

$$\mathrm{SS}_T = \left(\frac{T_{\mathrm{I}} - T_{\mathrm{I}}^2}{N} \right) \left(\frac{N^2}{T_{\mathrm{I}}(N - T_{\mathrm{I}})} \right) + (\quad)(\quad) + (\quad)(\quad)$$

$$= T_{\mathrm{I}} \left(1 - \frac{T_{\mathrm{I}}}{N} \right) \left(\frac{N^2}{T_{\mathrm{I}}(N - T_{\mathrm{I}})} \right) + (\quad)(\quad) + (\quad)(\quad)$$

$$= T_{\mathrm{I}} \left(\frac{N - T_{\mathrm{I}}}{N} \right) \left(\frac{N^2}{T_{\mathrm{I}}(N - T_{\mathrm{I}})} \right) + (\quad)(\quad) + (\quad)(\quad)$$

$$= N + N + N$$

$$= N(\mathrm{nca}) = 192(3) = 576.0$$

The fourth step is to calculate the OA column sums of squares. The sums of squares for each class are computed the same as two-class attribute data. The column sums of squares are multiplied by the class weight and added together to get the complete sums of squares:

$$\mathrm{SS}_{A_i} = \frac{(A_{1_i} - A_{2_i})^2}{N}$$

where A_{1_i} = cumulative occurrences for the first level of factor A in class i.

$$SS_A = \sum_{i=1}^{nca} SS_{A_i} W_i$$

Referring to the original factorial assignment of the OA,

$$SS_{A_I} = \frac{(88 - 88)^2}{192} = 0.0$$

$$SS_{A_{II}} = \frac{(90 - 94)^2}{192} = 0.0833$$

$$SS_{A_{III}} = \frac{(94 - 96)^2}{192} = 0.0208$$

$$SS_A = 0.0(13.09) + 0.0833(25.04) + 0.0208(97.01) = 4.1000$$

The remaining columns are calculated in a similar fashion.

The degrees of freedom in multiple-class analysis are similar to the other methods. However, each attribute class allows a comparison of the first and second levels of each OA column, which means each class analyzed provides a degree of freedom. Therefore, each column in the array has a similar calculation to factor A

$$v_A' = v_A \, (nca) = 1 \, (3) = 3$$

The total degrees of freedom is

$$v_T' = v_T \, (nca) = (N - 1) \, (nca) = 191(3) = 573$$

The ANOVA summary is shown in Table 5-63. In this instance, the

TABLE 5-63 ANOVA Summary Using Multiple Classes

Source	SS	v	V	F
A	4.100	3	1.370	1.370
B	9.360	3	3.120	3.120++
C	0.790	3	0.260	
D	3.180	3	1.060	
$A{\times}B/C{\times}D$	5.000	3	1.667	1.667
$A{\times}C/B{\times}D$	0.790	3	0.260	
$A{\times}D/B{\times}C$	1.090	3	0.360	
e^*	551.690	552	1.000	
T	576.000	573		

*e from classes used for F test
+ at least 90% confidence
++ at least 95% confidence
# at least 99% confidence

conclusions from the accumulation analysis are nearly the same as the two-class analysis. However, as demonstrated by this case study, when the frequency of occurrence is predominantly one of the end classes, the accumulation analysis doesn't offer any advantage over a simpler two-class analysis. Accumulation analysis works best when the distribution of frequency of occurrence is spread over many classes.

5-19-3   Case III: two-class attribute data; known occurrences only

In this situation, the frequency of occurrences can be treated the same as in Case I when the frequency of occurrences is used as data. Examples of this situation would be the number of surface imperfections per part for a material-handling system, the number of defects per painted panel, the number of knits per unit area of carpet, etc. The data would be the number of occurrences within a given trial.

5-20   Summary

This chapter started with the very simplest ANOVA that can be done and progressed sequentially through more complex ANOVAs to three-way ANOVA. This analysis method can be applied to more complex experimental situations or to any set of observations that is structured (has controlled groups with identical operating conditions) to allow decomposition of variation into accountable sources. The ANOVA method was then applied to designed experiments using orthogonal arrays. This chapter also discussed the accommodations necessary to complete an ANOVA on modified OAs having a mixture of two-, three-, or four-level factors including the dummy treatment, combination, and idle column methods.

This chapter covered the typical situations encountered in using attribute date. An experiment using attribute data is structured identically to variable data experiments. The analysis of the results is the fundamental difference between attribute and variable data experiments.

The next major step in the DOE process is to conduct a confirmation experiment to validate the conclusions drawn from the analysis and interpretation. The detail preparation and techniques used in confirmation experiments are covered in the next chapter.

Problems

5-1 Perform no-way ANOVA on these data points:

$$\text{Data } (y): 10, 7, 9, 12, 11, 8, 9$$

5-2 Perform one-way ANOVA (method 2) for four types of cardboard; the data is the burst load.

$$A_1 \quad 12, 15, 14$$
$$A_2 \quad 19, 20, 18, 21, 22$$
$$A_3 \quad 11, 11, 12, 12$$
$$A_4 \quad 23, 24$$

5-3 Perform two-way ANOVA on the data arranged for factors A and B.

$$A_1B_1 \quad 15, 13$$
$$A_1B_2 \quad 15, 14$$
$$A_2B_1 \quad 11, 12$$
$$A_2B_2 \quad 17, 18$$

5-4 Columns 2, 4, and 6 are merged to provide a four-level arrangement for factor A. Calculate SS_A for the data indicated.

Trial no.	1	2	3	4	5	6	7	8
Factor A level	1	2	4	3	1	2	4	3
y data	3	6	10	8	4	5	9	8

5-5 Calculate SS_B and SS_e for the dummy treatment data.

Trial no.	1	2	3	4	5	6	7	8
Factor B level	1	2	2'	3	1	2	2'	3
y data	3	0	1	6	4	1	1	7

5-6 Decompose polynomially this experimental data.

Trial no.	1	2	3	4	5	6	7	8
Factor C level	1	1	2	2	3	3	4	4
y data	4	5	2	1	6	6	10	9

5-7 If factors A to G are assigned to an L8 OA and the data for each trial is (two classes), what would be the ANOVA results?

Trial no.	1	2	3	4	5	6	7	8
Class 1	3	4	3	2	7	8	7	9
Class 2	7	6	7	8	3	2	3	1

5-8 If factors A to G are assigned to an L8 OA and the data for each trial is (four classes), what would be the ANOVA results?

Trial no.	1	2	3	4	5	6	7	8
Class 1	2	1	2	2	4	4	4	5
Class 2		1	3	2			1	
Class 3	2	2		1		1		
Class 4	1	1			1			

6

Confirmation Experiment

6-1 Introduction to Confirmation Experiment

A confirmation experiment is the final step in the first iteration of the DOE process, as discussed in Sec. 2-1-4. A confirmation experiment is performed by conducting a test using a specific combination of the factors and levels previously evaluated. The sample size of the confirmation experiment is larger than the sample size of any specific trial in the previous factorial experiment.

The key task is the determination of the preferred combination of the levels of the factors indicated to be significant by the analytical methods. The insignificant factors may be set at any desirable levels. Several samples are then generated under these constant conditions to observe the results. Depending on how close the average of these results is to the predicted or expected result, the experimenter then determines the next course of action.

The purpose of the confirmation experiment is to validate the conclusions drawn during the analysis phase. This is particularly important when screening, low-resolution, small fractional-factorial experiments are utilized. Because of the confounding within the columns, the conclusions should be considered preliminary until validated by a confirmation experiment. When a small fractional-factorial OA experiment is used and several factors contribute to the variation observed, it is likely that the best combination of factors and levels was not present in the OA test combinations. The confirmation experiment also serves the purpose of testing that specific combination of factors and levels.

When full-factorial experiments are conducted, a confirmation experiment is not as critical since there is no confounding present and less opportunity for misinterpretation; however, a confirmation experiment is still recommended because of the larger sample size used to evaluate that particular combination.

In product or process development work, the first experiment conducted generally does not provide the optimum performance and may not satisfy the experimental objective(s). Further testing with significant factors may lead to increased performance which eventually does meet the stated objective(s). A confirmation experiment should be conducted between each round of factorial experimentation to verify that the experimenter is drawing the correct conclusions and is following a trail leading to increased performance of the product or process. A flowchart, described in Sec. 6-6, provides a guide to the various decisions that may face the experimenter.

The steps in conducting a confirmation experiment are:

1.. Determine the preferred combination of the levels of the factors and interactions indicated to be significant by the analysis.

2. Determine the preferred levels for the factors indicated to be insignificant by the analysis.

3. Calculate the estimated mean for the preferred combination of the levels of significant factors and interactions.

4. Calculate the estimated standard deviation for the preferred combination of significant factors and interactions.

5. Determine the sample size for the confirmation experiment.

6. Calculate the confidence interval value.

7. Calculate the confidence interval for the true mean around the estimated mean.

8. Conduct tests under specified conditions.

9. Compare the confirmation test average result to the confidence interval for the true mean.

10. Determine the next course of action (use flowchart).

Refer to the analysis and interpretation methods, Chap. 5, to determine the significant and insignificant factors and interactions. The preferred levels for the significant factors will depend upon the quality characteristic being lower-is-better, nominal-is-best, or higher-is-better.

The preferred levels of the insignificant factors depend mainly on cost considerations. If there is a cost difference between any of the levels, then the most economic level should be used in the confirmation experiment. If there is no cost difference between any of the levels, then any level may be used in the confirmation experiment. Also, the levels may be considered to be upper and lower specification limits for this factor with respect to the quality characteristic being measured.

6-2   Estimating the Mean

Typically, the experimenter would like to obtain some particular response from a product or process: a higher average response is better (HB), a nominal value is best (NB), or a lower average response is better (LB). Depending on the characteristic, different treatment combinations will be chosen to obtain satisfactory results.

When an experiment has been conducted and the optimum treatment condition within the experiment determined, one of two possibilities exists:

1. The prescribed combination of factor levels is identical to one of those in the experiment.
2. The prescribed combination of factor levels was not included in the experiment (the lower the resolution, smaller fraction of a full-factorial experiment, the more likely this is to occur).

If situation 1 exists, one direct way to estimate the mean for that treatment condition is to average all the results for the trials which are set at those particular levels. If situation 2 exists, then a more indirect route will have to be taken to predict the average for that treatment condition. This method may also be used in situation 1, which will utilize more of the data to estimate the average.

The procedure depends upon the additivity of the factorial effects. If one factorial effect can be added to another to accurately predict the result, then good additivity exists. If an interaction exists, then the additivity between those factors is poor. The additivity of an interaction and other noninteracting factors, however, may be good.

A plot of two noninteracting factors is shown in Fig. 6-1. Geometrically, the midpoint on the B_1 line represents $\overline{B}_1$, the average of all the data under the B_1 condition. The same applies for the B_2 condition. $\overline{A}_1$ may be found midway between the B_1 and B_2 conditions when factor A is at the first level. The same applies for $\overline{A}_2$. If these four points are connected by two line segments as shown, $\overline{A}_1-\overline{A}_2$ and $\overline{B}_1-\overline{B}_2$, then the intersection represents $\overline{T}$, the average of the entire experimental results.

Assume the A_2B_2 treatment condition is to be estimated. Then $(\overline{A}_2-\overline{T})$ represents the A_2 effect to change the average from $\overline{T}$ to $\overline{A}_2$, and $(\overline{B}_2-\overline{T})$ represents the B_2 effect to change the average from $\overline{T}$ to $\overline{B}_2$. Since there is no interaction, the additivity is good. Then the estimate of the mean is

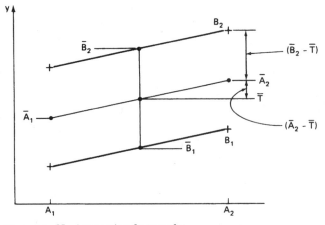

Figure 6-1 Noninteracting factors plot.

$$\hat{\mu}_{A_2B_2} = \overline{T} + (\overline{A}_2 - \overline{T}) + (\overline{B}_2 - \overline{T}) = \overline{A}_2 + \overline{B}_2 - \overline{T}$$

This is a very simple situation, but any number of factors can be combined if their additivity is good. The coefficient on the $\overline{T}$ term is one less than the number of items added to estimate the mean. In this case, two items were added, so the coefficient is equal to one. Another example provides an estimate of the $B_2C_1F_2G_2$ condition:

$$\hat{\mu}_{B_2C_1F_2G_2} = \overline{B}_2 + \overline{C}_1 + \overline{F}_2 + \overline{G}_2 - 3\overline{T}$$

If the characteristic is an HB characteristic and:

$$\overline{B}_1 = 1.30 \qquad \overline{B}_2 = 1.50$$
$$\overline{C}_1 = 2.30 \qquad \overline{C}_2 = 0.50$$
$$\overline{F}_1 = 0.90 \qquad \overline{F}_2 = 1.90$$
$$\overline{G}_1 = 1.00 \qquad \overline{G}_2 = 1.80$$
$$\overline{T} = 1.40$$

$$\hat{\mu}_{B_2C_1F_2G_2} = 1.50 + 2.30 + 1.90 + 1.80 - 3(1.40)$$
$$= 3.30$$

Note that the average of the $B_2C_1F_2G_2$ condition is higher than any of the individual factor averages because of the additivity of the effects.

When an interaction exists, the additivity is poor between the factors, especially when the interaction effect is larger than any of the main effects. To obtain the best estimate of a mean when an interaction is present, the trials that include that specific treatment condition should be averaged. Referring to Fig. 5.17, the hardness level is an HB characteristic, so the condition A_2B_2 is the best in the experiment. The effect of the A_2B_2 condition to change the average from $\overline{T}$ is represented by

$$\overline{A_2B_2} - \overline{T}$$

When two factors are interacting then

$$\overline{A_2} + \overline{B_2} - \overline{T} = \overline{A_2B_2} \tag{6-1}$$

The interaction effect is the difference between the left and right sides of Eq. 6-1.

If two other factor levels, C_1 and D_1, happen to also improve the hardness, how can the total effect be estimated? By considering the interaction as one item which has good additivity to other noninteracting items, an estimate may be made.

$$\hat{\mu}_{A_2B_2C_1D_1} = \overline{A_2B_2} + \overline{C_1} + \overline{D_1} - 2\overline{T}$$

Here, the poor additivity of the interacting factors is avoided when other noninteracting items are added. If

$$\overline{A_2B_2} = 9.5 \qquad \overline{C_1} = 7.2 \qquad \overline{D_1} = 7.5 \qquad \overline{T} = 6.875$$

then

$$\hat{\mu}_{A_2B_2C_1D_1} = 9.5 + 7.2 + 7.5 - 2(6.875) = 10.45$$

Again, any number of items can be combined when the additivity is good and the coefficient on $\overline{T}$ is 1 less than the number of items combined.

6-3 Confidence Interval Around the Estimated Mean

The estimate of the mean $\hat{\mu}$ is only a point estimate based on the averages of results obtained from the experiment. Statistically this provides a 50% chance of the true average being greater than and a 50% chance of the true average being less than $\hat{\mu}$. The experimenter would prefer to have a range of values within which the true average would be expected to fall with some confidence. The confidence interval is a maximum and minimum value between which the true average should fall at some stated percentage of confidence.

Confidence, in the statistical sense, means there is some chance of a mistake. For instance, the confidence that a number of 1, 2, 3, 4, or 5 could be rolled on a standard die is 5/6 or 83%. It is possible to roll a 6, so there is some risk of being wrong. When stating a confidence value for a confidence interval, experimenters are simply stacking the odds in their favor that the true average will fall between the stated limits. A high confidence may be chosen to reduce risk, but a wider confidence interval will result, lowering the chance of the true average being outside the stated limits.

There are three different types of confidence intervals (CIs) that Taguchi uses, depending on the purpose of the estimate.

1. Around the average for a particular treatment condition in the existing experiment

2. Around the estimated average of a treatment condition predicted from the experiment

3. Around the estimated average of a treatment condition used in a confirmation experiment to verify predictions

6-3-1 CI_1 for existing experimental treatment condition

This method of calculating a CI is the traditional statistical approach.

$$CI_1 = \sqrt{\frac{F_{\alpha;1;v_2} V_{ep}}{n}}$$

where $F_{\alpha;1;v_2}$ = F ratio required for:
$\quad\quad\quad\alpha$ = risk
$\quad$ Confidence = $1-$risk
$\quad\quad\quad v_1$ = 1
$\quad\quad\quad v_2$ = degrees of freedom for pooled error = V_{ep}
$\quad\quad\quad V_{ep}$ = pooled error variance
$\quad\quad\quad n$ = number of tests under that condition

The F ratio is determined from the same F tables used in ANOVA. The 1 degree of freedom for the numerator associated with the mean that is being estimated will always be a value of 1 for a confidence interval. The degrees of freedom for the denominator are the degrees of freedom v_{ep} associated with the pooled error variance V_{ep} of the experiment.

The CI is used in this manner:

$$\mu_{A_1} = \overline{A}_1 \pm CI_1$$

or

$$A_1 - CI_1 < \mu_{A_1} < A_1 + CI_1$$

Both of these statements are made at some chosen confidence level.

As an example of the use of a CI, the data for Fig. 5-16 has a 90% confidence interval of

$$CI_1 = \sqrt{\frac{F_{\alpha;1;v_e} V_{ep}}{n}}$$

where $F_{.10;1;9}$ = F ratio for 90% confidence = 3.36
   V_{ep} = .052
   n = 4

$$CI_1 = \sqrt{\frac{3.36(.052)}{4}} = .21$$

$$\overline{A}_1 = 1.050 \qquad \overline{A}_2 = 1.475 \qquad \overline{A}_3 = 2.050$$

$$1.050 - .21 < \mu_{A_1} < 1.050 + .21$$
$$0.840 < \mu_{A_1} < 1.260$$
$$1.265 < \mu_{A_2} < 1.685$$
$$1.840 < \mu_{A_3} < 2.260$$

As shown in Fig. 6-2, the CIs do not overlap each other at the 90% confidence level indicating, with at least 90% confidence, that the

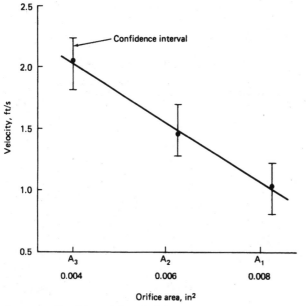

Figure 6-2   Confidence interval plot.

averages are different from each other. Also, a straight line passes through all of the CIs, indicating that the performance does not have a nonlinear component of any significance statistically. This method is complementary to polynomial decomposition for interpretation of product or process performance.

6-3-2   CI$_2$ for predicted treatment condition

The CI calculation is modified:

$$CI_2 = \sqrt{\frac{F_{\alpha;1;v_e} V_{ep}}{n_{eff}}}$$

$$n_{eff} = \frac{N}{1 + [\text{total degrees of freedom associated with items used in } \hat{\mu} \text{ estimate}]}$$

The effective sample size n_{eff} is applied to the treatment condition being estimated. For example, if an L16 had been used for the experiment with one test (repetition) per trial and

$$\hat{\mu}_{A_2B_2C_1D_2} = \overline{A_2B_2} + \overline{C}_1 + \overline{D}_2 - 2\overline{T}$$

then $N = 16$ and

$$n_{eff} = \frac{16}{(1 + 1 + 1 + 1)} = 4$$

One will note that a recommended four-factor assignment to an L16 does not have the $A_2B_2C_1D_1$ combination. The Taguchi method of using n_{eff} gives credit for the data points used in calculating $\overline{A_2B_2}$, $\overline{C}_1$, and $\overline{D}_2$. Again,

$$\hat{\mu}_{A_2B_2C_1D_2} = (\overline{A_2B_2} + \overline{C}_1 + \overline{D}_2 - 2\overline{T}) \pm CI_2$$

Another example of the n_{eff} calculation is

$$\hat{\mu}_{A_3B_3C_2} = \overline{A}_3 + \overline{B}_3 + \overline{C}_2 - 2\overline{T}$$

$$v_A = 2 \qquad v_B = 2 \qquad v_C = 1$$

$$n_{eff} = \frac{N}{(1 + 2 + 2 + 1)} = \frac{N}{6}$$

6-3-3 CI₃ for predicting a confirmation experiment

A confirmation experiment is used to verify that the factors and levels chosen from an experiment cause a product or process to behave in a certain fashion. A selected number of tests are run under constant, specified conditions to observe results that, the experimenter hopes, are close to the predicted value.

The difference between CI_2 and CI_3 is that CI_2 is for the entire population, i.e., all parts ever made under the specified conditions, and CI_3 is for only a sample group made under the specified conditions. Because of the smaller sample size relative to the entire population, CI_3 must be slightly wider and is modified.

$$CI_3 = \sqrt{F_{\alpha;1;v_e} V_{ep}[(1/n_{eff}) + (1/r)]}$$

r = sample size for the confirmation experiment, $r \neq 0$

As r approaches infinity (i.e., the entire population), the value $1/r$ approaches 0 and $CI_2 = CI_3$. As r approaches 1, the CI becomes wider.

The confirmation experiment is highly recommended to verify the experimental conclusions and is interpreted in this manner. If the average of the results of the confirmation experiment is within the limits of the confidence interval, then the experimenter believes that the significant factors as well as the appropriate levels for obtaining the desired result were properly chosen. If the average of the results of the confirmation experiment is outside the limits of the CI, then the experimenter has selected the wrong factors and/or levels to control the results at a desired value or has excessive measurement error, necessitating further experimentation. Other factors and other levels will have to be included in further rounds of experimentation. The confirmation experiment is the last step in an investigation to verify the understanding of what makes the product or process function properly. The confirmation experiment decisions are described more fully in Sec. 6-6.

6-4 Transformation of Percentage Data

The omega transformation of data is used in another situation where poor additivity can happen. That is when the data is in percentage values such as percent yield, percent loss, or percent defective. As these values approach 100% or 0%, the additivity may be poor; when the value is calculated, a value greater than 100% or less than 0% can be obtained.

To use the omega conversion method:

1. Convert data percent values to db values using the omega tables or formula.

2. Use $\hat{\mu}$ equation to estimate the mean with substituted omega values.

3. Convert the obtained db value back to the percent value using the omega tables or formula.

The omega tables are in Table D-7 of App. D.

In a fully saturated resolution 1 experiment, several factors may contribute something to reducing the percent defectives in a process. One or two of the factors probably contribute the most, but any help in reducing loss is useful. For example (percent loss is an LB characteristic),

$$\overline{A}_1 = 5\% \quad \overline{C}_1 = 10\% \quad \overline{D}_1 = 3\% \quad \overline{F}_1 = 8\%$$
$$\overline{A}_2 = 9\% \quad \overline{C}_2 = 4\% \quad \overline{D}_2 = 11\% \quad \overline{F}_2 = 6\%$$
$$\overline{T} = 7\%$$

Then

$$\hat{\mu}_{A_1C_2D_1F_2} = A_1 + C_2 + D_1 + F_2 - 3\overline{T}$$
$$= 5 + 4 + 3 + 6 - 3(7) = -3\%$$

Apply the omega transformation from App. D.

%	db
5	−12.787
4	−13.801
3	−15.096
6	−11.949
7	−11.233
1	−19.955

$$\hat{\mu}_{A_1C_2D_1F_2} = -12.787 - 13.801 - 15.096 - 11.949 + 3(11.233)$$
$$= -19.934 \text{ db}$$

Converting the db value back to percentage makes the estimate equal to approximately 1 percent. The omega transformation formula is

$$(db) = 10 \log \left[\frac{p}{(1 - p)} \right]$$

where p = decimal fractions $(0<p<1)$

The omega transformation converts fractions between 0 and 1 to values between minus infinity and plus infinity. This transformation is most useful when percentage values are very small or very large. When percentages are between 20 and 80%, the additivity is generally good.

6-5 Capability Estimates

In production process experimental development situations, process capability estimates may be made from the analytical information. The pooled error variance is the variation that is left over from all the pooled (statistically insignificant) factors, unknown (and uncontrolled) factors, and measurement error. These are the same things that will be contributing to the process variation when the significant factors are being controlled to specific conditions.

Error variance is, of course, equal to the square of the standard deviation of error. Using the square root of error variance will provide an estimate of the process standard deviation. Various capability indices may then be calculated, such as C_p and C_{pk}, depending on what quality methodology the experimenter is following. The tolerance limits must be known for these calculations to be performed.

The C_p value is a measure of how much wider the tolerance range is compared to the process limits. The formula is

$$C_p = \frac{(USL - LSL)}{6\,S_e}$$

where USL = upper specification limit
LSL = lower specification limit
S_e = pooled error standard deviation

Obviously, a higher value for C_p is better from a viewpoint of decreasing the quantity of discrepant parts manufactured over time. The loss function may be applied to determine the effectiveness of reducing variation and improving the process capability. The C_p value, of course, does not take into account the centering of the process around the nominal value of the quality characteristic. The C_p value reflects the potential of the process if the average were adjusted to be equal to the nominal dimension.

The C_{pk} value is a measure of how much wider the tolerance range is compared to the process limits, but takes the location of center of the process into account. The C_{pk} value is the minimum of

$$C_{pk} = \frac{(\text{USL} - \hat{\mu})}{3 \, S_e}$$

or

$$C_{pk} = \frac{(\hat{\mu} - \text{LSL})}{3 \, S_e}$$

where $\hat{\mu}$ = estimated average of the process (see Sec. 6-2)

The C_p and C_{pk} will be equal if the process is centered. C_{pk} will always be less than C_p if the process is off-center relative to the middle of the tolerance (nominal dimension).

These capability estimates are based on the pooled error variance of the whole experiment. This estimate may be an inflated value compared to the variation actually present under the preferred combination of significant factors and levels. There may actually be less variation present under more optimum combinations than other conditions used in the entire experiment. The variance of the parts actually created in the confirmation experiment should be calculated and compared statistically (F test) to the pooled error variance to determine if variation is really reduced. If so, the capability indices should be recalculated to show the improvement. The F ratio for the comparison would be

$$F = \frac{V_{ep}}{V_{\text{confirmation experiment sample}}}$$

when v_{ep} = pooled error variance degrees of freedom ·
$v_{\text{confirmation experiment sample}}$ = $r - 1$
r = sample size of the confirmation experiment

Chapter 7, "Parameter Design," will cover an additional method to estimate the variation for the confirmation experiment.

6-6   Confirmation Experiment Decisions

The experimenter is faced with several possibilities when the results of the confirmation experiment are finally available. A flowchart is a good method of depicting how these decisions relate to each other.

6-6-1   Confirmation experiment flowchart

A flowchart of the decisions made and actions taken after a confirmation experiment has been conducted is shown in Fig. 6-3. A successful

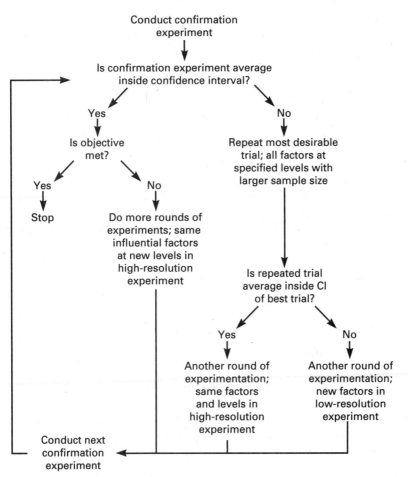

Figure 6-3 Confirmation experiment flowchart.

confirmation experiment (ce) is defined as one where the average of the samples $\bar{y}_{ce}$ falls within the predicted confidence interval for the true mean. Confidence interval number three may be used for this situation.

When $\bar{y}_{ce}$ falls within the confidence interval, the experimenter has evidence that the factors and interactions used in the estimate of the mean are, in fact, controlling the result. The next decision concerns whether the experimental objective(s) were met. If so, the experimental process may stop for all practical purposes, since a satisfactory solution to the problem has been identified. If the results do not meet the objective, then the significant factors and interactions should be evaluated at new levels to try to further improve the product or process. At this point, a high-resolution experiment may be used, since the few important factors and interactions have been identified. More details of conducting further experimentation are offered in Sec. 6-6-2. Of course, another confirmation experiment should be conducted after this round of experimentation.

When $\bar{y}_{ce}$ does not fall within the confidence interval of the expected results, then there has been some form of misinterpretation of the significant factors and interactions in the columns of the orthogonal array. This assumes that the measurement system is repeatable and reproducible. If $\bar{y}_{ce}$ does not fall within the confidence interval, there are two possible explanations. First, there is a confounded interaction that was not recognized in the previous analysis. When a column effect is large, the lowest-order item is assumed to be causing the difference; this is usually a factor assigned to that column. However, the difference could really be due to a confounded interaction in that column. The second possibility is that there are other unknown, uncontrolled factors which actually cause the variation observed in the experiment and the data fell into a pattern which made certain columns have large effects. To identify which of these two possible causes, interactions or other factors, made the confirmation experiment unsuccessful, the next step is recommended.

The most desirable trial results, from a technical viewpoint, should be identified. The factor levels in this trial determine the condition to be tested in, effectively, another confirmation experiment. A new confidence interval is calculated around the average of the results in the best trial and the average of the larger sample compared to the trial confidence interval. If the best trial $\bar{y}_{ce}$ falls within the expected confidence interval, then the experimenter has evidence that there was a confounded interaction which was not considered that caused the original confirmation experiment failure. When the factors and levels were reset to the best trial conditions, the results repeated the best results once again; this indicates that the true controlling factors are in the experiment but an interaction exists which caused misinterpretation. More experimentation should be done at a high resolution to identify which of those factors and interactions affect the results. If the best trial $\bar{y}_{ce}$ does not fall within the best trial confidence interval, then this indicates that there are factors which have an effect on the quality characteristic other than the factors previously evaluated. More experimentation will have to be performed with new factors in a low-resolution experiment to improve the product or process performance.

The confirmation experiment is the experimenter's safety net, especially when low-resolution experiments are utilized. Once again, low-resolution experiments are not necessarily discouraged since the few key factors and interactions may be identified with a minimum number of tests. The confirmation experiment provides a way to validate the conclusions when a smaller experiment is used in spite of the confounding that will be present.

6-6-2   Secondary rounds of experimentation

The initial screening experiment is intended to detect the factors which produce the desired effect. However, within the levels that were chosen for the experiment, the optimum results may not have been achieved and further experimentation may be necessary to move toward the optimum. Further experimentation should be done using significant factors from the screening experiment but with new levels that are anticipated to improve performance.

Depending on the type of characteristic (LB, NB, or HB), different strategies have to be used. Figure 6-4 shows the possibilities that might exist in a single-factor situation or a two-factor situation.

If one factor possesses the ability to influence the product or process response, then achieving a higher, nominal, or lower value is

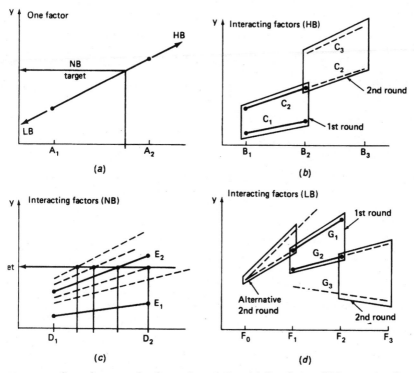

Figure 6-4   Secondary rounds of experimentation. (*a*) One factor; (*b*) interacting factors (HB); (*c*) interacting factors (NB); (*d*) interacting factors (LB).

relatively easy. Appropriately extrapolating or interpolating along the line of best fit to the results will allow the experimenter to use a certain level of the factor of concern to obtain a certain result (refer to graph a of Fig. 6-4).

Interactions between factors are not as easily addressed and depend on the type of characteristic, as indicated in graphs b, c, and d in Fig. 6-4. The HB and LB characteristics are very similar in approach. New levels for the factors of concern are chosen as a way to extrapolate into yet uninvestigated territory. The extrapolated lines need the verification of the secondary round of experimentation. This round of experimentation should include one of the levels of the primary round as a way to verify the repeatability of the experiment. If the average of the duplicated condition of the second round is within the confidence interval of the first round, then this is an indicator, but not a guarantee, that the controlling factors are in the experiment.

If some of the experimental results are higher than the target value and some are lower for an NB characteristic, then several combinations of the two factors will provide the desired nominal value. Economics relevant to these factors and their levels should be considered in choosing the factor levels. If the target value is greater than all of the interaction treatment condition averages, then the same approach as shown for an HB characteristic should be used. If the target value is less than all of the interaction treatment condition averages, then the same approach as shown for an LB characteristic should be used.

6-7   Example Experiments

6-7-1   Water pump experiment

The confirmation experiment for the water pump experiment used the significant factors of gasket design and pump finish with levels of new design and smooth finish, respectively. These factors were indicated to be the significant factors in the observation method and ANOVA methods. The column effects table shows that the second level for gasket design, new design, and the second level for pump finish, smooth finish, had the lowest sums for leak rating, an LB characteristic.

The insignificant factors may be set at either level to validate all the conclusions from this experiment. The production front cover was used since it was a less expensive design, no sealant was used since that would add expense with no improvement, the front-to-back torque sequence was used since that was preferred by production people, and the low torque values were used for both bolts. In this case, the low and high levels for torque were considered to be valid specification limits from a leakage viewpoint. Perhaps, from a loosening due to vibration or temperature cycling viewpoint, the specifications would have to be set at different values.

Ten pumps were built with the aforementioned combination of factors and levels and tested for leakage. The expectation at the time was that none of the pumps should leak. That expectation was reinforced by the two trials with no leak results that had new gasket design and smooth pump finish as factor levels.

The calculations for the estimated mean, confidence interval value, and the confidence interval for the true mean are as follows. Referring to the column effects table for the water pump, Table 5-3, the significant factor preferred level averages are

$$\overline{B}_{2,\text{ new design}} = \frac{2}{4} = 0.5$$

$$\overline{E}_{2,\text{ smooth finish}} = \frac{5}{4} = 1.25$$

$$\overline{T} = \frac{15}{8} = 1.875$$

$$\hat{\mu}_{B_2 E_2} = \overline{B}_2 + \overline{E}_2 - \overline{T} = 0.5 + 1.25 - 1.875 = -0.125$$

Referring to the ANOVA summary table for the water pump leak experiment, Table 5-41, $V_e = 0.125$ with $v_e = 5$. The CI_3 will be equal to

$$CI_3 = \sqrt{F_{\alpha;1;v_e} V_{ep}[(1/n_{\text{eff}}) + (1/r)]}$$

$$F_{.10;1;5} = 4.06$$

$$n_{\text{eff}} = \frac{N}{(1 + v_{\text{items in estimate}})} = \frac{8}{(1+2)} = 2.67$$

$$r = 10$$

$$CI_3 = \sqrt{4.06(0.125)\left[\left(\frac{1}{2.67}\right) + \left(\frac{1}{10}\right)\right]} = 0.49$$

$$\hat{\mu} - CI < \mu < \hat{\mu} + CI$$

$$-0.125 - 0.49 < \mu < -0.125 + 0.49$$

$$-0.615 < \mu < 0.365$$

The true mean confidence interval includes the value of zero leak rating which was the result desired for the confirmation experiment. Also, the variation under this condition is expected to be zero; all pump assemblies should not leak.

The 10 pump assemblies were tested and found to perform satisfactorily with no leakage. According to the confirmation experiment flowchart, Fig. 6-3, the results were inside the confidence interval and met the objective of eliminating leaks, so experimentation was discontinued at this point. Further development work was necessary, however, to determine the degree of "smoothness" required to perform properly and to establish meaningful and measurable tolerance limits for surface finish. The economic impact of the new gasket design was justified since that factor had the largest effect to eliminate leakage.

6-7-2   Die-cast piston experiment

The mean may be estimated, a confidence interval around the mean calculated, and process capability estimated for the die-cast piston hardness experiment using previous ANOVA summary information, Table 5-43, plus the following information.

Grand average of experiment: $\overline{T} = 71.19$

Level means for significant factors and interactions:

	A(%Copper)	B(%Magnesium)
Level 1 mean	69.87	72.29
Level 2 mean	72.50	70.08

D(Water) $\times$ E(Air) interaction level combinations:

$$\overline{D_1E_1} = 73.00 \qquad \overline{D_2E_1} = 75.75 \qquad \overline{D_1E_2} = 64.42 \qquad \overline{D_2E_2} = 71.58$$

In an attempt to obtain the highest possible hardness for improved piston durability, the most desirable combinations of factors and interaction combinations are those with the greatest average. The water cooling and air cooling interaction combination has good additivity with the copper and magnesium effects. An estimate of the mean is

$$\hat{\mu}_{A_2B_1D_2E_1} = \overline{A}_2 + \overline{B}_1 + \overline{D_2E_1} - 2\overline{T}$$

$$\hat{\mu}_{A_2B_1D_2E_1} = 72.50 + 72.29 + 75.75 - 2(71.19)$$

$$\hat{\mu}_{A_2B_1D_2E_1} = 78.16 \, R_B$$

Using a 90% confidence level and sample size of 20 pistons, the confidence interval value is

$$F_{.10;1;42} = 2.83$$

$$V_{ep} = 9.022$$

$$n = 20$$

$$n_{eff} = \frac{48}{(1 + 3)} = 12$$

$r = 40$ (two hardness measurements for each piston)

$$CI_3 = \sqrt{2.83(9.022)\left[\left(\frac{1}{12}\right)+\left(\frac{1}{40}\right)\right]} = 1.66$$

The confidence interval for the confirmation experiment is

$$\hat{\mu} - CI < \mu < \hat{\mu} + CI$$

$$78.16 - 1.66 < \mu < .78.16 + 1.66$$

$$76.50 < \mu < 79.82$$

The experimenter is 90% confident that the average of the 20 parts made in the confirmation experiment will fall within this interval. In trial 5, there are three pistons, six hardness values, representing the preceding combination of factor levels. The average hardness of trial 5 is 77.50 R_B which, even with a small sample, is inside the confidence interval. The results from trial 5 provide a preliminary indicator that the confirmation experiment should succeed.

Based on the complete experimental variation and current specification limits of 68 to 78 R_B, the standard deviation and process capability estimate is

$$V_e = 9.022 \qquad S_e = \sqrt{V_e} = 3.004$$

$$C_p = \frac{(USL - LSL)}{(6\,S_e)}$$

$$C_p = \frac{(78-68)}{(6 \times 3.004)} = \frac{10}{18.02} = 0.55$$

This is not considered an acceptable C_p value in quality control work. C_p values less than 1 indicate that the process is not capable and will make a substantial number of parts outside the specification limits.

Based on variation within the best trial, the standard deviation and process capability may be estimated again. The standard deviation for trial 5, the best results in the complete experiment, is 0.837 hardness points. Therefore,

$$S_5 = 0.837$$

$$C_p = \frac{10}{6\,(0.837)} = \frac{10}{5.022} = 1.99$$

This capability value is considerably higher than the previous calculation. The first capability estimate included variation from all of the tested combinations of factors and levels plus positional variation. The second estimate is for the best combination, even though it includes the positional variation. The standard deviation of the 20 confirmation experiment parts should be calculated to verify that the actual variation is indeed this low.

The C_{pk} value in this case will be a negative number, since the estimated average is slightly higher than the upper specification limit. Obviously, if the increased hardness offered by controlling all four factors is to be utilized, the specification limits would have to be changed with the customer's approval. If the specification limits could not be changed, then other combinations of the four factors could be estimated to obtain a result within the tolerance range.

6-7-3 Casting cracks confirmation experiment (see Sec. 5-19-1)

A confirmation experiment was proposed with 300 castings poured at the higher temperature and left in the mold a shorter amount of time. Since this is an attribute data situation, the sample size will need to be large to prove that the defective rate is truly low. The other two factors, cooling fan and shot blast, were set at the most economical conditions (fan on for faster cooling and subsequent material handling with one pass of shot blast). All 300 castings were defect-free, making a total of 348 castings which were defect-free for this process condition. Based on the binomial distribution, this statistically provided a confidence of 97% that the defective rate was less than 1%.

$$P(\text{defects} = 0) = (1 - p)^n = 1 - C$$

where P = probability
p = percent defective
n = sample size
C = confidence level

Because of this mathematical relationship, it is impossible to say at 100% confidence that the defective rate is 0%; only a very large sample size will allow this position to be approached. Still, this was a substantial improvement over recent performance and may really be 0% defective; it is just statistically impossible to prove.

An additional 1200 castings were magnafluxed as production parts were poured; all of these were also defect-free (now 97.9% confidence that the percent defective is less than 0.25%). At this point, magnaflux was discontinued and the casting process allowed to perform at the 100% yield conditions with respect to casting cracks.

6-8 Summary

This chapter addressed the final step in one iteration of the DOE process in order to validate the conclusions drawn from the earlier analysis. The confirmation experiment should always be performed, especially when small fractional-factorial experiments are utilized. The price the experimenter pays for the smaller experiment size and cost is an increased amount of confounding expressed as a low-resolution experiment. To overcome the possible misinterpretation of a low-resolution experiment, a confirmation experiment is performed.

A flowchart provides guidance for the experimenter when the confirmation experiment is completed. There are several situations and subsequent decisions that confront the experimenter after the confirmation tests are run. This flowchart shows why the DOE process is an iterative one: only one outcome allows the experimenter to stop after one round of experimentation. All the other situations require further experimentation of some form.

Problems

6-1 How would the equation be written to calculate an estimate of the mean for the condition $A_2B_1E_2F_2$?

6-2 If $\overline{A}_2 = 55$, $\overline{B}_1 = 53$, $\overline{E}_2 = 50$, $\overline{F}_2 = 42$, and $\overline{T} = 35$, what will be the value for the estimated mean of Prob. 6-1?

6-3 What is average of the D_2E_2 interaction combination for the die-cast piston experiment in Table 4-8?

6-4 What F ratio is used in a confidence interval calculation if a 90% confidence interval is desired and v_{ep} from the ANOVA summary table equaled 12?

6-5 Using the die-cast piston data from Table 4-8, a 90% confidence value $V_{ep} = 9.022$ and $v_e = 42$, what is the confidence interval for the $\overline{D}_2$ treatment condition (use CI_1)?

6-6 Using the information in Sec. 6-7-2, what would be the confidence interval for the estimated mean using the CI_2 value?

7

Parameter Design

7-1 Parameter and Tolerance Design
Explanation

This chapter concerns probably the largest contribution to quality methodology that Taguchi has made. The designed experiments addressed in all of the previous chapters but the first have existed since the 1930s. However, previous experimental approaches looked upon all factors as causes of variation. If these causes could be well controlled or eliminated, then product or process variation would be reduced, and, therefore, quality would be improved. But if a product is sensitive to ambient temperature variations, how can anyone control or eliminate temperature in a customer's environment? The answer is obvious; ambient temperature variations can neither be controlled nor eliminated without a large expense. Ambient temperature in a factory might be controlled very well to eliminate the temperature effect on a machine's performance, but the world's atmosphere cannot be controlled, so all kinds of devices are exposed to large temperature variations. Therefore, a different approach is required if product quality is to be improved. This approach was entitled parameter design by Taguchi and is covered in the majority of this chapter. Parameter design is used to improve quality without controlling or eliminating causes of variation. Controlling or eliminating causes of variation may be expensive compared to a parameter design approach.

Taguchi views the design of a product or process as a three-phase program:

1. System design
2. Parameter design
3. Tolerance design

System design is the phase when new concepts, ideas, methods, etc., are generated to provide new or improved products to customers. One way to remain competitive in the world economy is to be a leader in utilizing technology. However, the technological advantage disappears quickly because it may be copied. If a competitor can fabricate the same new idea in a more uniform manner, then the technological advantage is more than lost. The parameter design phase is crucial to improving the uniformity of a product and can be done at no cost or even at a savings. This means that certain parameters of a product or process design are set to make the performance less sensitive to causes of variation. The tolerance design phase improves quality at a minimal cost. Quality is improved by tightening tolerances on product or process parameters to reduce the performance variation. This is done only after parameter design.

Typically, when a problem is detected in product development, an engineer may jump directly to tolerance design; when tolerances are tightened, variation will be reduced and quality improved. However, tightening tolerances may be expensive and completely unnecessary if parameter design were used first. One serious mistake a designer can make is to use expensive materials, components, or processes for a product when lower-cost items may be used if a parameter design approach is applied.

7-2   Control and Noise Factors

To begin the discussion on parameter design, one must consider kinds of design and development factors. Taguchi separates factors into two main groups: control factors and noise factors, as mentioned in Sec. 2-3-5. Control factors are those which are set by the manufacturer and cannot be directly changed by the customer. An engine manufacturer can dictate the material for the pistons, the tension on the piston rings, the piston-bore clearance, etc., which cannot be easily modified by the customer (of course, many hot-rod fanatics do this very thing to improve short-term output horsepower, usually at the sacrifice of durability). Noise factors are those over which the manufacturer has no direct control but which vary with the customer's environment and usage. In general, noise factors are among those which the manufacturer desires not to have to control at all.

Noise factors can be placed in three categories:

1. Outer noise

2. Inner noise

3. Product noise

Outer noises are environmental factors such as ambient temperature, humidity, pressure, or people. Even different batches of materials may be viewed as outer noise to a production process. Inner noises are function- and time-related, such as deterioration, wear, fade of color, shrinkage, and drying out. Outer noises produce variation from outside the product; inner noises produce variation from inside or within the product; and product noise manifests itself in part-to-part variation. Products may have sensitivity to all three forms of noise simultaneously. The design quality of a product or process provides less functional variation due to outer or inner noise. Production quality provides less functional variation from one part to another and close to a target value. Taguchi refers to design quality efforts as off-line quality control (QC) and production quality efforts as on-line quality control. Off-line QC is any of the design and development activities that may take place before products are manufactured and available to customers. When products are manufactured to be available for customers, the on-line activities begin. Of course, the on-line activity must be well planned before the start of production. The loss function reflects the off- and on-line QC efforts in the equation:

$$L(y) = k[S_y^2 + (\bar{y} - m)^2]$$

The variance S_y^2 is reduced and the target value m determined during the off-line phase. The average value produced $\bar{y}$ is controlled during the on-line phase. Both of these phases will reduce the loss associated with the product. The real purpose of a quality control chart is to provide the proper, close-to-the-target, average value from a process. Incidental (special) causes of variation may be identified and controlled or eliminated, but this is the tolerance design approach. If a large amount of variation is present at the introduction of a process, this is an indication of the lack of off-line QC efforts. The more off-line QC work done, the more robust a process or product is against disturbances (outer and inner noise) in the environment and life of a product. Again, to reduce loss to society in the loss function there must be reduced variance (off-line QC work) and an average near the target value (on-line QC work).

Parameter and tolerance design take on additional meanings with the loss function approach. Parameter design is used to dampen the effect of noise (reduce variance) by choosing the proper level for control factors. Parameter design is used to improve quality without controlling or removing the cause of variation, to make the product robust against noise factors. Tolerance design is used to reduce the variation by tightening tolerances around the chosen target value of the control factors. Tolerance design reduces or eliminates the effect of causes of variation. By using parameter and tolerance design, the true critical characteristics (control factors) can be identified and minimized in number. Because of the principles of parameter and tolerance design, major emphasis is placed on true control factors and very little emphasis on true noise factors. Primarily, noise factors are used in experiments to expose the robust levels of the control factors.

In traditional statistical terms, control and noise factors are also generally categorized as fixed and random factors, respectively. Fixed factors are those that may have specific values selected for their levels in a designed experiment, while random factors are those that have levels which are randomly determined, such as different operators, different machines, etc. The parameter design concept still applies, however the factors are described.

Parameter design is not a difficult concept to understand or apply. The scientist, engineer, or technician most likely just needs to know how to perform parameter design in order to obtain the benefits.

7-3 Introduction to Parameter Design

Imagine that an experiment is structured like that shown in Table 7-1, with control factors only assigned to the L8 OA. Why bother to assign noise factors when they cannot be controlled or eliminated without some expense? Three repetitions for each trial are planned in this instance. What will happen as data points y_1, y_2, and y_3 are collected for each trial (not necessarily one after another; remember the randomization strategies)? Will y_1 be identical to y_2, and identical to y_3? Noise effects, all those uncontrolled and unknown factors that were actually varying from one degree to another as the experiment was run, cause

TABLE 7-1   Simple Parameter Design Experiment

	Control factors							Data		
	A	B	C	D	E	F	G			
	Column no.									
Trial no.	1	2	3	4	5	6	7	y_1	y_2	y_3
1	1	1	1	1	1	1	1	*	*	*
2	1	1	1	2	2	2	2	*	*	*
3	1	2	2	1	1	2	2	*	*	*
4	1	2	2	2	2	1	1	*	*	*
5	2	1	2	1	2	1	2	*	*	*
6	2	1	2	2	1	2	1	*	*	*
7	2	2	1	1	2	2	1	*	*	*
8	2	2	1	2	1	1	2	*	*	*

*Data points.

the differences from data point to data point. Rather than assuming that error variation is an aggregate of all these noise effects and equally distributed in all treatment conditions, parameter design utilizes these repetitions to aid in identifying what levels of what control factors might have reduced variation. There may be some control factor(s) for which one of the levels is more robust against the noise effects.

In the previous example, repetitions were used to assess the noise effect on some performance characteristic(s) of interest. However, these noise effects were merely accidental as the experiment was being conducted; error variance is all the unknown and uncontrolled factors which are varying in unknown amounts. A better parameter design strategy would be to force noise effects into the experiment in a different manner. Table 7-2 shows an experimental layout with an inner array for control factors only and an outer array for noise factors only. If these factors are all mixed in an inner array, then this is a traditional cause detection experiment. If a product is found to be sensitive to ambient temperature, what then? Can instructions be supplied with the product to tell the customer to use the product only within a certain temperature range? Not many people would buy such a product.

This parameter design strategy separates the control factors from the noise factors by using inner and outer arrays, respectively. Now noise factors (X, Y, and Z) such as temperature and operators could be

TABLE 7-2 Inner/Outer OA Parameter Design Experiment

								L4 OA outer array (noise factors)			
							Z	1	2	2	1
							Y	1	2	1	2
							X	1	1	2	2
	L8 OA inner array (control factors)										
	A	B	C	D	E	F	G				
	Column no.							Data			
Trial no.	1	2	3	4	5	6	7	y_1	y_2	y_3	y_4
1	1	1	1	1	1	1	1	*	*	*	*
2	1	1	1	2	2	2	2	*	*	*	*
3	1	2	2	1	1	2	2	*	*	*	*
4	1	2	2	2	2	1	1	*	*	*	*
5	2	1	2	1	2	1	2	*	*	*	*
6	2	1	2	2	1	2	1	*	*	*	*
7	2	2	1	1	2	2	1	*	*	*	*
8	2	2	1	2	1	1	2	*	*	*	*

*Data points.

assigned to the outer array to find some level of a control factor that doesn't have much variation in the results in spite of the noise factors definitely being present.

In this experimental arrangement, there are 32 separate test conditions. A trial number specifies a test condition with respect to control factors, but the outer array specifies four test conditions with respect to noise factors for that trial. If tests are very expensive to run, a thorough outer array may be precluded and only one noise factor thought to be strong (or noise factors combined into the best and worst conditions) may be used.

Also, if product or process performance can be modeled mathematically, then a much less expensive experiment can be run. The equation, i.e., mathematical model, of a system will have many terms with various powers and magnitudes of coefficients for those terms. The designed experiment can use the terms as factors and substitute into the equation high and low values as prescribed by the OA. Obviously, repetitions are not necessary, since the equation will give exactly the same result each time. The performance of electronic devices, voltage, amperage, etc., can be readily modeled mathematically, which makes the electronics field a prime candidate for designed experiments using simulation. In addition, electronic devices abound with nonlinear performance characteristics, making a parameter design approach very desirable. The nonlinear aspect of parameter design will be discussed later in Sec. 7-6-1. Hardware should be used on a few of the test conditions to verify the mathematical model.

7-4   Signal-to-Noise Ratios

The control factors that may contribute to reduced variation (improved quality) can be quickly identified by looking at the amount of variation present as a response. All past analyses in this text have addressed which factors might affect the average response, but now there is interest in the effect on variation as well. Taguchi has created a transformation of the repetition data to another value which is a measure of the variation present. The transformation is the signal-to-noise (S/N) ratio. The S/N ratio consolidates several repetitions (at least two data points are required) into one value that reflects the amount of variation present. There are several S/N ratios available depending on the type of characteristic; lower is better (LB), nominal is best (NB), or higher is better (HB).

7-4-1   Different *S/N* ratios

The S/N ratio, which condenses the multiple data points within a trial, depends on the type of characteristic being evaluated. The equations for calculating S/N ratios for LB, NB, or HB characteristics are:

1. Lower is better.

$$S/N_{LB} = -10 \log\left(\frac{1}{r}\sum_{i=1}^{r} y_i^2\right)$$

where r = number of tests in a trial (number of repetitions regardless of noise levels).

2. Nominal is best.

$$S/N_{NB_1} = -10 \log V_e \quad \text{(variance only)}$$

$$S/N_{NB_2} = +10 \log\left(\frac{V_m - V_e}{rV_e}\right) \quad \text{(mean and variance)}$$

3. Higher is better.

$$S/N_{HB} = -10 \log\left(\frac{1}{r}\sum_{i=1}^{r}\frac{1}{y_i^2}\right)$$

The LB and HB S/N ratios are both easy to calculate; each repetition is entered into the equation. However, the NB S/N ratio needs further explanation. Both ratios contain the value V_e, and NB_2 contains V_m. These values are determined by doing a no-way ANOVA (see Sec. 5-7) on all the repetitions for a trial. Even though the no-way ANOVA method appeared to have no real purpose in the analysis discussion, the S/N ratio utilizes this decomposition approach. Recall that the variance due to the mean V_m has 1 degree of freedom associated with it always; therefore,

$$V_m = \frac{SS_m}{v_m} = \frac{SS_m}{1} = SS_m = r(\bar{y})^2$$

$$V_e = \frac{SS_e}{v_e} = \left(\frac{SS_T - SS_m}{r - 1}\right)$$

$$SS_T = \sum_{i=1}^{r} y_i^2$$

$$S/N_{NB_1} = -10 \log\left(\frac{SS_T - SS_m}{r-1}\right)$$

$$S/N_{NB_2} = +10 \log\left(\frac{SS_m - V_e}{rV_e}\right)$$

The S/N for NB_1 is a function of variation only and the S/N for NB_2 is a function of both average and variation. For this reason, the use of

S/N_{NB_1} is recommended by Hunter for the NB type of characteristic.* The log V_e also provides a tendency toward normality for the variance statistic. The S/N_{NB_1} might be considered as a substitute for LB or HB characteristics since log V_e is independent of the average but S/N_{LB} and S/N_{HB} are not. The following example demonstrates the effect of using the two different S/N_{NB} ratios.

7-4-2 Example *S/N* calculation

Four possible trial results are shown which have differing amounts of variation in an experiment arranged with an inner array and a two-level noise factor array, N_1 and N_2.

Trial 1 results:

$$N_1 \; 11 \; 10 \; 9 \qquad N_2 \; 11 \; 10 \; 9$$

$$SS_T = \sum_{i=1}^{r} y_i^2 = 604$$

$$SS_m = r(\bar{y})^2 = 6(10)^2 = 600$$

$$V_e = \left(\frac{SS_T - SS_m}{r-1} \right) = \frac{(604-600)}{5} = .8$$

$$S/N_{\text{NB}_1} = -10 \log V_e = -10 \log .8 = .97 \text{ db}$$

$$S/N_{\text{NB}_2} = +10 \log \left(\frac{SS_m - V_e}{rV_e} \right) = 10 \log \left[\frac{600-.8}{6(.8)} \right] = 20.96 \text{ db}$$

Trial 2 results:

$$N_1 \; 12 \; 10 \; 8 \qquad N_2 \; 12 \; 10 \; 8$$

$$S/N_{\text{NB}_1} = -5.05 \text{ db}$$

$$S/N_{\text{NB}_2} = 14.93 \text{ db}$$

Note that both S/N ratios dropped virtually the same amount of dbs due to the increase in variability; the average remained exactly the same. As variation is doubled, the S/N ratio drops approximately 3 db.

Trial 3 results:

$$N_1 \; 12 \; 11 \; 10 \qquad N_2 \; 10 \; 9 \; 8$$

---

*J. Stuart Hunter, "Signal-to-Noise Ratio Debated," *Quality Progress,* May 1987, pp. 7–9.

$$S/N_{NB_1} = -3.02 \text{ db}$$
$$S/N_{NB_2} = 16.98 \text{ db}$$

Note that the S/N ratios are not as high as in trial 1, but are higher than in trial 2. The average value is exactly the same, but variation is more than in trial 1 and less than in trial 2.

Trial 4 results:

$$N_1 \ 21 \ 20 \ 19 \qquad N_2 \ 21 \ 20 \ 19$$
$$S/N_{NB_1} = -.97 \text{ db}$$
$$S/N_{NB_2} = 26.99 \text{ db}$$

In this case, the variation is the same as in trial 1, but the average has increased. One can see that S/N_{NB_1} is a function of the variability only and S/N_{NB_2} is a function of both variability and average. Table 7-3 summarizes the results of this example, in which it has been demonstrated that reduced variation relative to the average causes an increase in the S/N value; S/N is always an HB characteristic (not the absolute value in the case of negative db values; -3.02 db is higher than -5.05 db).

7-4-3   Equivalent variance quality characteristic

Occasionally, difficulty in comprehending the meaning of the different S/N ratios is encountered and, because of this problem, an alternative measure of variation within a trial is offered. The S/N_{NB_1} may be converted to another measure of variation which is much easier to comprehend.

$$S/N_{NB_1} = -10 \log V_e = -10 \log S_e^2 = -20 \log S_e$$

The value S_e is the standard deviation within a trial; the variation of all repetitions for which almost all technical people have an appreciation. The base 10 log of standard deviation is used as a traditional statistical

TABLE 7-3   Nominal Is Best *S/N* Ratios

Trial no.	Data		S/N	
	N_1	N_2	N_1	N_2
1	11, 10, 9	11, 10, 9	0.97	20.96
2	12, 10, 8	12, 10, 8	−5.05	14.93
3	12, 11, 10	10, 9, 8	−3.02	16.98
4	21, 20, 19	21, 20, 19	0.97	26.99

transformation to make a normal distribution out of the skewed standard deviation distribution. To simplify this even further, the coefficient in front of the log S value (-20) may be disregarded and just the log S value used as a quality characteristic. The -10 coefficient in the S/N calculation is used to magnify the numbers for easier analysis and to make the S/N value a higher-is-better characteristic. Therefore, log S is inversely proportional to -10 log V. The S/N ratios are set up to make a higher value more desirable. As a comparison, the lower the log S value, the more desirable the results within a trial.

7-4-4   Analysis of S/N ratio as a response

The S/N ratio (or log S value) is treated as a response of the experiment, which is a measure of the variation within a trial when noise factors are present. If an outer array is used, the noise variation is forced in an experiment; with pure repetitions (no outer array), the noise variation is unforced. S/N ratio is a response which consolidates repetitions and the effect of noise levels into one data point. A standard ANOVA can be done on the S/N ratio which will identify factors significant to increasing the average value of S/N and subsequently reducing variation.

7-5   Parameter Design Strategy

When the ANOVA on the raw data (identifies control factors which affect average) and the S/N data (identifies control factors which affect variation) are completed, the control factors may be put into four classes:

Class I: Factors which affect both average and variation (significant in both ANOVAs)

Class II: Factors which affect variation only (significant in S/N ANOVA only)

Class III: Factors which affect average only (significant in raw data ANOVA only)

Class IV: Factors which affect nothing (not significant in both ANOVAs)

The parameter design strategy is to select the proper levels of classes I and II to reduce variation and class III to adjust the average to the target value. Class IV may be set at the most economical level since nothing is affected. Figure 7-1 shows example plots for the four classes of factors. The fundamental parameter design approach is to move all but one design (control) parameter to a region of low-

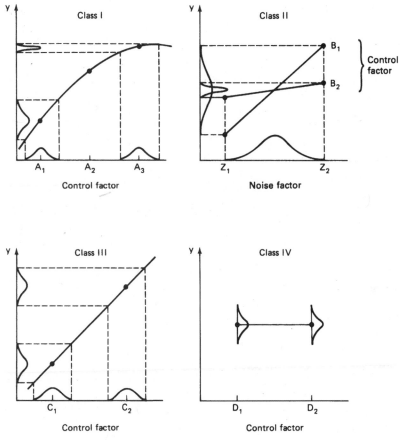

Figure 7-1 Classes of control factors.

response slope to make those parameters insensitive to variation. The remaining factor(s) should be linear (adjustment) factors to obtain the appropriate average response (quality characteristic). When starting with two-level experiments, insignificant factors are already parameter designed and significant factors may be investigated further to discover nonlinear properties, if possible.

7-6   Case Studies of Parameter Design

Two case studies to be discussed use the property of the class I control factor primarily to reduce variation. Both of these case studies require the use of a class III factor to adjust the average, but the case study on heat treatment is an interesting application. The third case study utilizes a class II factor for parameter design.

7-6-1   Case study: electronic circuit*

A power supply circuit is required to provide a certain output voltage to another electronic circuit. A designed experiment has been done on all the electronic components to determine some significant factors in the various classes. This experiment could have been done by modeling the circuit mathematically and calculating the output voltage when components take on different levels rather than testing physical hardware. Again, electronic circuitry is particularly adaptable to this approach, and a mathematical model should be used whenever possible to reduce test costs and time.

Imagine that two components are strongly related to output voltage performance as determined by a designed experiment. The voltage could be a function of transistor gain, component A, which was evaluated at three levels of gain and resistance, and component B, which was evaluated at two levels as shown in Fig. 7-2. The curves labeled B_1 and B_2 are generated from the average voltages obtained under the six possible treatment conditions of factors A and B. Based on this performance, a typical designer might specify a transistor gain of A_x to provide the targeted average voltage. However, since transistors vary in gain from one part to another, the output voltage will vary from assembly to assembly due to transistor gain variation as indicated by distribution I.

A parameter design approach, in contrast, would be to specify the transistor gain in the neighborhood of A_z. At this value of gain, the variation of transistors could be even greater, but the variation of output voltage would be substantially reduced as indicated by distribution II. The problem here is that the average voltage produced is greater than the target value, so an "adjustment" factor is required. Resistance may be altered to the value B_w to reduce the average voltage to the target value. In this case, the transistor gain is a class I factor and resistance is a class III factor.

Several advantages are gained by the parameter design approach. First, larger variation of transistor gain is not passed on to the variation in output voltage. This allows lower-quality components to be used without lowering the quality of the overall circuit. This makes the circuit robust against transistor product noise. Making a circuit robust against one noise makes it robust against other noises also. If transistor gain changes slightly with operating temperature, an outer noise, then this will not be passed on to variation in output voltage. If transistor gain changes slightly with age, an inner noise, this will not

---

*Genichi Taguchi and Yu-in Wu, *Off-Line Quality Control,* Central Japan Quality Control Association, Nagaya, 1979, pp. 30–31.

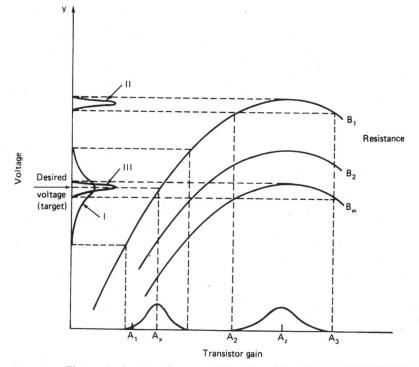

Figure 7-2 Electronic circuit performance. (*Reproduced from Genichi Taguchi and Yu-in Wu*, Off-Line Quality Control, *Central Japan Quality Control Association, Nagaya, 1979, p. 30. Used by permission.*)

be passed on either. Second, because the sensitivity of transistor gain is greatly reduced, the number of critical components in the circuit is reduced. Using the parameter design approach minimizes the number of critical factors in a product or process, which subsequently reduces the cost of the on-line QC efforts.

At this point, tolerance design should be applied if variation due to transistor gain and/or resistance causes excessive variation in output voltage. Tolerance design is simply reducing the variation in a component to reduce variation in the system. The improved quality required of some of the components will increase the cost of the system. The loss function should be used to justify any cost increases due to higher-quality components.

Now it should be apparent why parameter design must be applied before tolerance design. If the original level of A_x were used as the target value for transistor gain and tolerance design applied first, then the component costs of both the transistor and resistor would increase. The tolerances on the transistor and resistor would have to be tightened to reduce variation in output voltage. However, now the circuit is still sensitive to outer and inner noise because of utilizing the high slope area of the performance curve.

Parameter design should be applied first to improve quality at no additional cost. Tolerance design should be applied second, if necessary, to improve quality at a minimal cost. The nonlinear performance characteristic is one that should be utilized to the fullest extent for low-cost quality improvement.

7-6-2   Case study: heat treatment

This case study involved the use of a nonlinear characteristic of a process. A heat treatment process, carbonitriding, was primarily used to create wear-resistant surfaces on an engine component. Unfortunately, a side effect of the heat treatment process caused a slight growth in height of the component. The unpredictability of this growth had been causing a scrap problem after the heat treatment operation. An experiment was designed to investigate various heat treatment factors for their effect on part growth.

Out of all those studied, the testers screened out one key factor which had a nonlinear tendency with respect to change in height (growth) during heat treatment. Three levels of the amount of ammonia (NH_3) introduced into the heat treatment atmosphere during a batch were evaluated with results shown in Fig. 7-3. The past production level had been at the 5-ft^3 (0.1415 m^3) value. The device which measured the amount of NH_3 introduced into the heat treatment atmosphere had a precision of approximately ± 1 ft^3 (0.0283 m^3). The variation in amount of ammonia caused substantial variation in batch-to-batch change in height.

The parameter design approach was utilized and a level of 8.75 ft^3 (0.2476 m^3) was chosen for the production conditions. Any value above 10 ft^3 (0.2830 m^3) caused adverse chemical reactions. At the new level, the measuring device doesn't have to be replaced or improved to cause an improvement in the quality of the heat treatment process. Only the average change in height had to be accommodated. The adjustment factor was easy to come by; only the preheat treatment height had to be changed to handle the increased amount of growth. The parts would have to be machined to a slightly smaller value than in the past, but the resultant part after heat treatment would be much more consistent from batch to batch.

Increasing the amount of NH_3 per batch did increase costs; NH_3 costs approximately $0.0025 per cubic foot ($0.0883 per cubic meter). For approximately one cent per batch, the quality of this process was vastly improved. The tolerance design approach of improving the NH_3 measuring device would have entailed a much greater cost than one cent per batch. The subsequently reduced inspection costs and scrap

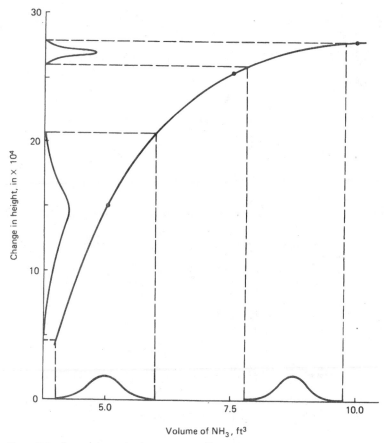

Figure 7-3 Ammonia content versus growth.

rate saved approximately \$20,000 annually for the amount of one cent per batch investment.

7-6-3 Case study: fishing reel line roller

Two fishing reels of identical design but different sizes were purchased from the same company. The reels were open-faced spinning reels of intermediate and large line capacity and operated identically with respect to the line retrieval mechanism. Line is wound around a stationary spool (which has an axis parallel to the fishing pole) by running the line over a roller that revolves around the spool on a bail mechanism. The roller axis of rotation is nominally perpendicular to the line spool and fishing pole axes but does not intersect the axis of the line spool. Figure 7-4 shows two views of the spool and roller

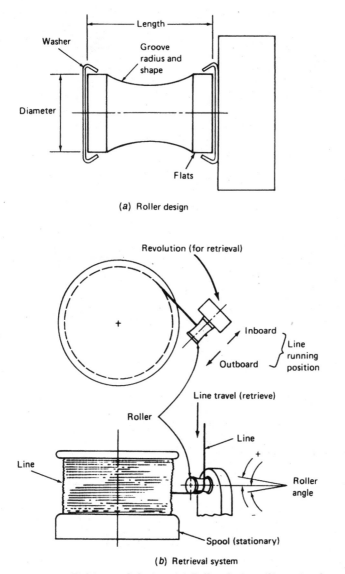

(*a*) Roller design

(*b*) Retrieval system

Figure 7-4 Fishing reel design. (*a*) Roller design; (*b*) retrieval system.

mechanism for line retrieval. The roller axis is nominally perpendicular to the spool axis, but one reel assembly, by chance, had a positive roller angle and the other, by chance, a negative roller angle. The intent is to have the line run in the center of the hourglass-shaped roller to prevent line deterioration through many retrieves. However, the roller with the positive angle would tend to allow the line to run

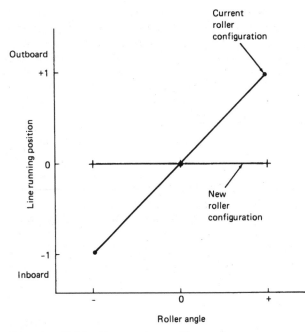

Figure 7-5   Fishing line running position performance.

on the outboard end against the washer and the roller with the negative angle would tend to allow the line to run on the inboard end against the washer. A plot of this performance is shown in Fig. 7-5 as the current roller design. The line running against the side washers would deteriorate very rapidly and perhaps under high load (perish the thought of it being a large steelhead trout) would separate. The reels were reworked, using shimming and filing, by the customer to achieve a zero roller angle, which provided a line retrieval position in the center of the roller. This again was a tolerance design approach (tightening tolerances to reduce performance variation), but it was the only alternative the customer had at this point.

A parameter design approach would treat the roller angle as a product noise factor (desiring not to have to control the roller angle). A tolerance design approach would be a mistake at this time; tightening tolerances on roller angle by adjustment, selective assembly, or scrap of excessively angled parts to eliminate the tendency to run off the end of the roller would add unnecessary cost. A parameter design approach would be to find a roller configuration that would be insensitive to roller angle with regard to line-running position.

A new high-quality roller configuration, as yet undetermined, is also shown in Fig. 7-5. The line would run in the center in spite of roller angle variation. This shows the class II type of control factor which interacts with a noise factor. In this case, the roller configuration factor(s) would be the control factor(s) and roller angle would be the noise factor. A recommended experiment to determine the appropriate roller configuration might include the factors shown in Table 7-4 and the experimental layout might look like the OA in Table 7-5.

In this proposed experiment, the inner array is a typical L8 OA and the outer array is a one-factor, three-level array. To conduct the experiment, the various roller configurations would have to be manufactured, the line retrieved under load, with the roller angle adjusted to the $-$, 0, and $+$ positions, and the running position of the line observed. A response value of $+1$ could be assigned if the line ran off the outboard end, a value of 0 assigned if the line ran in the center, and a -1 assigned if the line ran off the inboard end. The data could then be analyzed by ANOVA of the S/N value for an NB characteristic.

This experimental approach would reveal if any control factor(s) would provide a roller design that would be more robust against roller angle variation without necessarily increasing cost. An angler using such a roller design would notice that the reel is robust against other

TABLE 7-4   Fishing Reel Parameter Design Factors and Levels

Roller factors		Level 1	Level 2
A	End flats	Yes	No
B	Groove radius	0.250 in (6.35 mm)	0.375 in (9.53 mm)
C	Groove shape	Radius	Vee
D	Diameter	0.250 in	0.375 in
E	Length	0.375 in	0.500 in (12.70 mm)

TABLE 7-5   Fishing Reel Parameter Design Experiment

	Roller factors							Roller angle		
	A	B	C	D	E	e	e			
	Column no.									
Trial no.	1	2	3	4	5	6	7	$-$	0	$+$
1	1	1	1	1	1	1	1	*	*	*
2	1	1	1	2	2	2	2	*	*	*
3	1	2	2	1	1	2	2	*	*	*
4	1	2	2	2	2	1	1	*	*	*
5	2	1	2	1	2	1	2	*	*	*
6	2	1	2	2	1	2	1	*	*	*
7	2	2	1	1	2	2	1	*	*	*
8	2	2	1	2	1	1	2	*	*	*

*Data points.

noises as well. Other noises are line diameter (strength), speed of retrieve, load on the line, amount of line currently on the spool, etc. This approach to reel design would provide a much higher quality reel at no cost.

7-7   Analysis of Inner/Outer Array Experiments

The ANOVA for an inner/outer OA experiment can be somewhat more complex than for an inner OA only because of the additional sources of variation. The outer array factors add sources of variation, but because of the structure of the inner/outer array, there are many additional two-factor interactions. Each inner array factor interaction with an outer array factor may be estimated; the inner/outer array structure is a full-factorial experiment for control and noise factors.

A recommended approach, however, is to perform an ANOVA on the raw data for inner array factor effects and temporarily ignore the outer array factor effects. This approach focuses on the control factors for having an effect on the mean (adjusting the average). A second ANOVA performed using the S/N ratio or log S transformed data identifies indirectly which control factors have an interaction with noise factors and, therefore, have the potential to reduce variation.

7-7-1   Case study: die-cast component

The second example experiment utilized throughout the text to this point makes a good example of analyzing an experiment with inner and outer array structures. The process factors are all control factors in the inner array, as shown previously in Table 4-8. The noise factor of position on the piston is assigned equivalently to the first column of an L4 outer array. Position has two levels, the dome and the skirt positions. There are effectively six repetitions within a trial for each of the eight control factor conditions. There are effectively three repetitions for each trial and noise factor condition. Note that only three pistons were measured for hardness in two different places; several castings were made under the actual conditions to let the process stabilize for the designated operating conditions.

7-7-2   ANOVA and interpretation of raw data

The first ANOVA focusing on the raw data and the inner array factors is identical to the analysis discussed in Sec. 5-16-2 and summarized in Table 5-43. That analysis indicated that the most important factors were air cooling, water cooling, copper %, and magnesium % in descending order of effect on hardness. There was also a very slight air cooling and water cooling interaction effect. These are the factors and interactions that have an effect to increase or decrease the average hardness of pistons, disregarding the position on the piston.

7-7-3 ANOVA and interpretation of *S/N* data

Analysis of the S/N ratio follows after the raw data from each trial has been transformed. In this case, the data is transformed to an HB S/N value which is contained in Table 7-6. An ANOVA is conducted on the S/N values as if they were one data point for each trial. The ANOVA table is summarized in Table 7-7 and in a pooled form in Table 7-8. This analysis indicates that air cooling, factor E, and water cooling, factor D, are also very important to reducing variation of piston hardness. This variation is in two forms; variation from piston to piston and within a piston are both reduced. The average S/N values for the levels of factors D and E are

$$\overline{D}_1 = 36.67 \qquad \overline{E}_1 = 37.42$$
$$\overline{D}_2 = 37.34 \qquad \overline{E}_2 = 36.59$$

The estimated mean is

$$\mu_{D_2E_1} = \overline{D}_2 + \overline{E}_1 - \overline{T} = 37.34 + 37.42 - 37.01 = 37.75$$

TABLE 7-6 Hardness
***S/N* Transformation**

Trial no.	S/N, db
1	37.26
2	37.04
3	35.61
4	37.38
5	37.78
6	36.56
7	37.14
8	37.26

TABLE 7-7 *S/N* ANOVA Summary

Source	SS	v	V
A	0.26	1	0.26
B	0.19	1	0.19
C	0.02	1	0.02
D	0.88	1	0.88
E	1.39	1	1.39
$A \times B/D \times E$	0.23	1	0.23
$A \times E/B \times D$	0.05	1	0.05
T	3.02	7	

TABLE 7-8 *S/N* Pooled ANOVA Summary

Source	SS	v	V	F	P
D	0.88	1	0.88	5.87+	24.17
E	1.39	1	1.39	9.27++	41.06
e_p	0.75	5	0.15		34.77
T	3.02	7			100.00

+ at least 90% confidence
++ at least 95% confidence
# at least 99% confidence

Note that this S/N value for the best combination of factors that reduces variation is very close to the values obtained in trials 4 and 5. These trials both have the D_2E_1 combination of levels.

7-7-4 ANOVA and interpretation of log *S* data

If log S values were calculated for each trial instead of the S/N_{HB} value, the eight trial results would appear as in Table 7-9. Treating this as a single repetition of a new quality characteristic, the unpooled ANOVA is summarized in Table 7-10 and pooled ANOVA is summarized in Table 7-11. In this analysis, factors D and E were both significant, which means they are both important to reduce variation. The relative importance is reversed, however, due to the log S quality characteristic versus the S/N quality characteristic.

The average log S values for the levels of factors D and E are

$$\overline{D}_1 = 0.455 \qquad \overline{E}_1 = 0.115$$

$$\overline{D}_2 = 0.088 \qquad \overline{E}_2 = 0.428$$

The estimated mean is

$$\mu_{D_2E_1} = \overline{D}_2 + \overline{E}_1 - \overline{T} = 0.088 + 0.115 - 0.271 = -0.068$$

TABLE 7-9 Log *S* Transformation of Hardness

Trial no.	Log S
1	0.278
2	0.068
3	0.842
4	0.102
5	−0.077
6	0.552
7	0.253
8	0.151

TABLE 7-10   Log S ANOVA Summary

Source	SS	v	V	F	P
A	0.0211	1	0.0211	—	3.46
B	0.0347	1	0.0347	—	5.68
C	0.0223	1	0.0223	—	3.66
D	0.2726	1	0.2726	—	44.66
E	0.1987	1	0.1987	—	32.55
$A{\times}B/D{\times}E$	0.0559	1	0.0559	—	9.16
$A{\times}E/B{\times}D$	0.0050	1	0.0050	—	0.83
T	0.6103	7	—	—	100.00

TABLE 7-11   Log S Pooled ANOVA Summary

Source	SS	v	V	F	P
D	0.2726	1	0.2726	9.80++	40.10
E	0.1987	1	0.1987	7.14++	27.99
e_p	0.1390	5	0.0280	—	31.91
T	0.6103	7	—	—	100.00

+ at least 90% confidence
++ at least 95% confidence
# at least 99% confidence

This value is slightly worse (slightly higher) than the best log S value (trial 5) obtained in the original experiment.

When the log S or S/N_{NB_1} is used as the quality characteristic, an estimate of standard deviation may be made for the optimum combination. Since

$$\log S = -0.068$$

$$S = 10^{-0.068} = 0.85\ R_B$$

The standard deviation of the best trial was 0.84 R_B. The standard deviation of the confirmation experiment should be close to the estimated value of 0.85 R_B. A traditional chi-square confidence interval around this value should be determined to decide upon the success or failure of the confirmation experiment.

The factors in the die-casting experiment can then be classified.

Factor	Affects average	Affects variability
A	(2)#	—
B	(1)++	—
D	(2)#	(2)++
E	(1)#	(1)++

+ at least 90% confidence
++ at least 95% confidence
# at least 99% confidence
NOTE: Numbers in parentheses indicate the best level

In this case, the same levels of two of the significant factors provide a higher average and reduced variability, so nothing has to be compromised. In some situations, the levels of factors which improve the average and improve uniformity may conflict, so a compromise may have to be reached. Also, a compromise may have to occur when multiple responses are considered and the same factor level may cause one response to improve and another to deteriorate.

7-7-5   Control and noise factor interactions

Since control factors D (water cooling) and E (air cooling) are both statistically significant in the S/N or $\log S$ analyses, this implies that there is a significant interaction with these factors and the noise factor Z (position on the piston). The average hardnesses R_B for the appropriate combinations are

$$\overline{D_1 Z_1} = 66.17 \quad \overline{E_1 Z_1} = 74.08$$
$$\overline{D_1 Z_2} = 71.25 \quad \overline{E_1 Z_2} = 74.67$$
$$\overline{D_2 Z_1} = 74.08 \quad \overline{E_2 Z_1} = 66.17$$
$$\overline{D_2 Z_2} = 73.25 \quad \overline{E_2 Z_2} = 69.83$$

The plots for these interactions are seen in Fig. 7-6. Even though the water cooling by position and air cooling by position interactions can be shown to be statistically significant in this case, position is a noise factor and the only option is to choose the proper levels for water and air cooling. Notice that the average hardness of D_2 and E_1 levels is higher and more consistent in spite of the noise factor. Using level 2 of factor D and level 1 of factor E is parameter design with respect to component hardness.

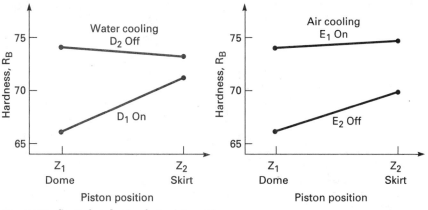

Figure 7-6   Control and noise factor interactions.

TABLE 7-12   ANOVA Summary with Noise Interactions

Source	SS	v	V	F	P
A	82.688	1	82.688	13.48#	5.62
B	58.500	1	58.500	9.54#	3.85
D	295.000	1	295.000	48.08#	21.22
E	487.688	1	487.688	79.49#	35.37
$A{\times}B/D{\times}E$	58.500	1	58.500	9.54#	3.85
$D{\times}Z$	105.020	1	105.020	17.12	7.26
$E{\times}Z$	28.520	1	28.520	4.65	1.64
e_p	245.398	40	6.135	—	21.19
T	1361.314	47	—	—	100.00

The ANOVA summary table could be redone to include these interactions and would appear as in Table 7-12. Other inner-outer array interactions, even though they may be present, are not useful in either changing the average hardness of the piston or reducing variation piston to piston or within a piston.

7-8   Alternative Inner/Outer OA Experiment

This approach to parameter design is the most sophisticated of the three strategies discussed so far. The first approach used repetitions only as a noise source, as in Table 7-1. A second and somewhat more comprehensive approach utilized an outer array to intentionally include noise in the experiment, as in Table 7-2. The third approach uses the same three-level array as the inner and outer array to investigate the presence of nonlinear performance characteristics. The first approach used no intentional noise factors, the second approach intentionally used outer noise factors, and the third approach intentionally uses product noise factors.

In new product or process designs the tendency for nonlinear performance may or may not be known, but a strategy to detect this phenomenon and allow parameter design to be done is desired. The nonlinear performance provides the opportunity to make the performance robust against product noises. Assume that a product has nonlinear performance with respect to design parameter C, as indicated in Fig. 7-7. Three levels of parameter C could detect the nonlinear tendency. However, if the levels of parameter C are held very precisely for the experiment, the S/N ratios may not differ very much. If product noise is forced into the experiment by using plus and minus values of parameter C around the nominal values of parameter C, then the S/N ratio would indicate a different variance. So each of the three nominal levels of parameter C (C_1, C_2, C_3) could have three levels within them ($C_{1'}$, $C_{1''}$, $C_{1'''}$, for instance). This experiment emulates the real world

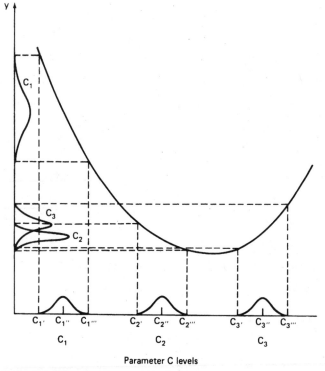

Figure 7-7 Three-level parameter design.

situation of having chosen a nominal value for parameter C, but, due to manufacturing variation, that value is not obtained exactly in every product that is produced. Sometimes the actual value of parameter C is a little lower than nominal and sometimes a little higher than nominal. Determining which nominal is best to reduce sensitivity to product variation is the key. In Fig. 7-7, the best of the three levels of parameter C is the nominal of C_2. Because of the nonlinear performance, the optimum level is between $C_{2'}$ and C_3. This level would accommodate the largest variation in parameter C without transmitting this variation through to the performance characteristic y.

To structure an experiment to take advantage of this strategy is an extension of the inner/outer OA approach. The three-level inner array is assigned only control factors as before, with all of the primes indicating the nominal values for the parameters to be investigated. The outer three-level array is used to designate the variation around the nominals of the control factors; level 1 might be 5% below the nominal; level 2 would be the actual nominal; and level 3 might be 5% above the nominal.

7-8-1   Case study: automobile hood hinge

The automobile hood hinge mechanism mentioned in Sec. 1-4 could have a parameter design study done to develop as consistently functioning a hood as possible. The force to close the hood from the open position, which is automatically held by the mechanism, is an NB characteristic as previously mentioned. The mechanism shown in basic form in Fig. 7-8 is a four-bar linkage with a coil spring located at the pivot point of links A and B. Links A and B have a mechanical stop to establish the angle between those links when the hood is in full open position. The design parameters that are available to investigate are the lengths of the four links (A, B, C, and D), the angle between links A and B, and the coil spring force. Link D could be set at the maximum length that would allow the hood to reach the closed position, since the experiment would include all combinations of the lengths of links A, B, and C. With five factors at three levels, an L18 OA is required for the experimental layout, so the resultant inner/outer array arrangement is as shown in Table 7-13.

The levels for the control factors in the inner array are shown in Table 7-14, and the levels for noise in the outer array are shown in Table 7-15. Obviously, this type of parameter design experiment requires many data points. In many situations, the cost and time to prepare hardware preclude this approach. For these reasons, this parameter design approach is used mainly when a mathematical simulation of the product is available. For instance, an equation might be written that would calculate the force required to close the hood

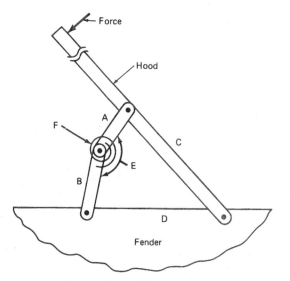

Figure 7-8 Automobile hood hinge mechanism.

TABLE 7-13   Three-Level OA for Parameter Design

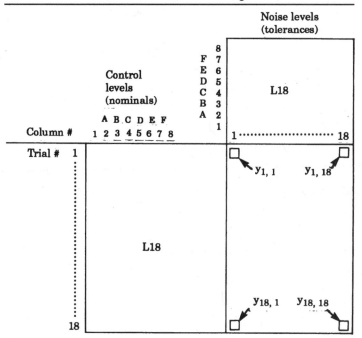

TABLE 7-14   Parameter Design Nominal Levels

Control factors	Nominal levels		
	1	2	3
A	7	9 in.	11
B	7	9	11
C	9	11	13
D		Dependent on A, B, and C	
E	150°	160°	170°
F	10	20 ft-lb	30

TABLE 7-15   Parameter Design Noise Levels

Control factors	Noise levels		
	1	2	3
A	−1%	Nominal	+ 1%
B	−1%	Nominal	+ 1%
C	−1%	Nominal	+ 1%
D	−1%	Nominal	+ 1%
E	−5%	Nominal	+ 5%
F	−10%	Nominal	+ 10%

based on the lengths of links A, B, C, and D; the angle E between links A and B; and the coil spring load F. The equation would read that the force y is a function of A, B, C, D, E, and F:

$$y = f(A, B, C, D, E, F)$$

A value could then be calculated for each of the 324 possible combinations of the control and noise levels. The value for the $y_{1,1}$ test (first trial in control array and first trial in noise array) would be:

$$y_{1,1} = f(A = 6.93, B = 6.93, C = 8.91, D = 8.91, E = 142.5, F = 9.0)$$

The formula would allow a force value to be predicted for each of the possible combinations of control and noise levels. Then an S/N ratio could be calculated to provide a measure of the variation caused by the noise levels. Two ANOVAs would then be completed to identify which control factors and levels affected the average force and the variation of the force. With this identification complete, the sensitivity of the hood closing force to variation in the hinge mechanism could be reduced. This, again, would make the hood closing force less sensitive to other sources of noise, such as amount of lubricant (outer noise) and wear in the mechanism (inner noise).

7-9   Dynamic Characteristics

A special type of quality characteristic is addressed in this section. Discussion up to this point has focused on variable and attribute characteristics; dynamic characteristics are a third grouping to consider. Dynamic characteristics are those characteristics related to any system having an input and an output or outputs. Many devices that we use every day fall into this form of characteristic, such as thermostat controls for home ambient temperature, heater control settings for automobile ambient temperature, steering wheel settings for automobile directional control, and measurement systems in general (scales, micrometers, etc.). A measurement system is shown in simplistic form in Fig. 7-9. The input (signal) is the item to be measured and the output is the value observed via the measurement system. The observed value deviates, hopefully a small amount, from the true value because of the noise effects entering the system. Parameter design, therefore, applies to dynamic characteristics and utilizes the S/N ratio as an indicator of the consistency of a given system.

The key property being sought in dynamic characteristic design is the linearity of any given device. For example, the automobile heater control mechanism should provide a linearly increasing temperature from the "coldest" position to the "hottest" position. If the mechanism

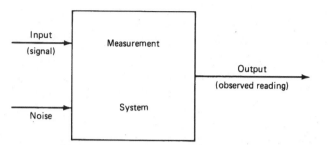

Figure 7-9   Measurement system.

is nonlinear, customers will have difficulty obtaining a personally comfortable temperature in the zones where the mechanism is very sensitive. An ideal heater control response is shown in Fig. 7-10 and has a purely linear response. A poor system response is shown in Fig. 7-11, which has three zones with different slopes. This will make it difficult to obtain comfortable settings when temperatures are near the extremes.

The outer array arrangement for dynamic characteristics is structured differently to accommodate the desire to have linear system responses. The signal factor is usually set up with three levels having

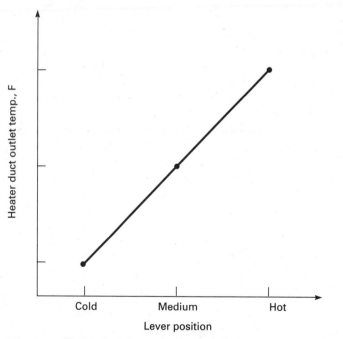

Figure 7-10   Heater control system response (ideal).

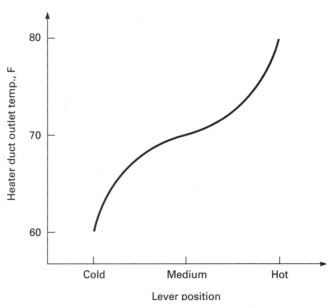

Figure 7-11   Heater control system response (poor).

equal increments between the levels. This allows the linearity of the system response to be evaluated. Noise factors usually have only two levels since that is all that is required to obtain the noise effects. Systems that are very linear over the operating range and are very insensitive to noise effects have utilized the parameter design approach. The heater control mechanism had three levels set up for the lever position which is the signal factor; this is the device the customer uses to obtain a given comfortable temperature. Noise factors that might have a significant effect on the temperature control system would be outside air temperature and fan speed. A fundamental structure for the outer array of a dynamic characteristic experiment is shown in Table 7-16. This generic arrangement uses L4 OAs, but any two-level arrangement will suffice if there is one strong noise factor.

7-9-1   Measurement system parameter design

Parameter design can be accomplished on measurement systems by using the signal-to-noise concept. The input or signal into a measurement system is some unknown, but true, value. The diameter of a piston bore is not actually known but provides the signal to a dial bore gauge. Also entering into the measurement system is noise, which may take many forms, such as variations in operators, temperature, and vibration. Noise is really any disturbance to the true purpose of

TABLE 7-16   Dynamic Characteristic Outer Array

Noise factor(s)	Signal factor		
	M_1	M_2	M_3
Z	1 2 2 1	1 2 2 1	1 2 2 1
Y	1 2 1 2	1 2 1 2	1 2 1 2
X	1 1 2 2	1 1 2 2	1 1 2 2

the measurement system, which is to translate the signal into some scalar value in a precise manner. The accuracy of a measurement system may require compensation or calibration, but the precision (repeatability) is the primary property sought in the device. The output of a measuring system is the observed value. The dial bore gauge mentioned previously converts the abstract value of diameter of the bore into a concrete length value of inches or millimeters. The true dimension of the bore is unknown, but the observed value indicated by the gauge will have to be assumed to be correct because the noise effect is also unknown. There are ways to experiment with measurement systems to assess the repeatability and the impact of noise which require repeated measurement of the same parts in the same position. The average of the repeated measurements tends to cancel out some of the noise effects and provide a better estimate of the true value of the item being measured.

The signal-to-noise concept for quantifying the ability of a measurement system to translate a signal into an observed value is quite useful. The S/N value ratios the power of the signal to the power of the noise as they appear in the output.

$$S/N = \frac{\text{power of the signal}}{\text{power of the noise}}$$

In a measurement system parameter design experiment, the strategy is to include the signal factor with noise in the outer array. Again, only control factors are assigned to the inner array.

Measurement system example 1.   In an automotive thrust washer test, one of the responses of interest is wear of the material which might allow components to shift in location due to a change in thickness of the washer. A gauge, similar in concept to a micrometer or height gauge, is used to detect a change in thickness after a particular thrust washer test. The gauge should be able to detect very small changes in thickness. Since the parts will have to be measured before the test and after the test, the precision of the measuring device will have to be very good. The signal factor will be some thickness, the levels of which should take on three equally spaced, known values to span the range of thickness values to be measured by the device. The known values of thickness should be a measurement standard such as gauge blocks or some other trusted standard. The noise factor in this experiment might be different operators used to collect test data. Control factors for the gauge would be such things as gauge-to-washer contact area, gauge-to-washer load, etc. A summary of the factors in the experiment is:

Control factors $A–G$

Noise factor N (operators 1 and 2)

Signal factor S (thickness standard)

The experimental arrays would be arranged like those in Table 7-17. Signal factor levels could be such values as 0.050 in (1.27 mm), 0.150 in (3.81 mm), and 0.250 in (6.35 mm), which span the typical values for automotive thrust washers. The recorded data values are the actual readings the measuring device provided when measuring the thickness of the signal level. These values are in thousandths of an inch.

To calculate the S/N value for each trial, the data is rearranged into a noise-versus-signal matrix. The first trial is represented in Table 7-18. With this arrangement, the total variation of these values can be decomposed into variance due to the signal (both linear and quadratic), the noise, the signal and noise interaction, and error. To be useful, however, a gauge should be as linear as possible, so a large amount of variation due to the linear component of the signal is desirable and all other variation should be considered as error. Therefore,

TABLE 7-17   Measurement System Parameter Design

								Outer array				
			Inner array					Signal level				
	A	B	C	D	E	F	G	1 (0.050)		2 (0.150)		3 (0.250)	
			Column no.										
Trial no.	1	2	3	4	5	6	7	N_1	N_2	N_1	N_2	N_1	N_2
1	1	1	1	1	1	1	1	45	53	155	145	248	253
2	1	1	1	2	2	2	2	35	43	145	155	263	258
3	1	2	2	1	1	2	2	48	52	149	151	252	248
4	1	2	2	2	2	1	1						
5	2	1	2	1	2	1	2						
6	2	1	2	2	1	2	1						
7	2	2	1	1	2	2	1						
8	2	2	1	2	1	1	2						

NOTE: First three trials for example only.

TABLE 7-18   Trial 1 Data Summary

Operator	Signal level			Totals
	1	2	3	
1	45	155	248	448
2	53	145	253	451
Totals	98	300	501	899

$$SS_T = SS_{S_1} + SS_{S_q} + SS_N + SS_{S \times N} + SS_e$$

$$= \sum_{i=1}^{r} y_i^2 - \frac{T^2}{r} = 45^2 + 53^2 + \dots + 253^2 - \frac{899^2}{6}$$

$$= 40697.0$$

where r = number of repetitions in a trial

The linear sum of squares (refer to the formula from Table D-5 of App. D and Sec. 5-17-2) for a three-level factor is

$$SS_{S_1} = \frac{(W_1 S_1 + W_2 S_2 + W_3 S_3)^2}{W_T R}$$

$$= \frac{[-1(98)+0(300)+1(501)]^2}{2(2)} = 40{,}602.0$$

where R = number of repetitions within the factor S levels

The ANOVA summary is shown in Table 7-19. The S/N calculation is based on the second nominal-is-best form

$$S/N_{NB_2} = 10 \log \left[\frac{V_m - V_e}{r V_e} \right]$$

TABLE 7-19   ANOVA Summary for Trial 1

Source		SS	ν	V
	S_1	40,602.	1	40,602.00
Error	$\left\{\begin{array}{c} S_q \\ N \\ S \times N \\ e \end{array}\right.$	95.	4	23.75
	T	40,697.	$\bar{5}$	

In a gauge study, the variance due to S_1 is substituted for the variance due to the mean, and R for r, and the equation then becomes

$$S/N_{\text{measurement}} = 10 \log\left[\frac{V_{S_1} - V_e}{RV_e}\right]$$

For trial 1, $S/N_1 = 10 \log\left[\dfrac{40{,}602 - 23.75}{2(23.75)}\right] = 29.32$

For trial 2, $SS_T = 49{,}157$ $SS_{S_1} = 49{,}062$ $SS_e = 95$

$$S/N_2 = 10 \log\left[\frac{49{,}062 - (95/4)}{2(95/4)}\right] = 30.14$$

For trial 3, $SS_T = 40{,}018$ $SS_{S_1} = 40{,}000$ $SS_e = 18$

$$S/N_3 = 10 \log\left[\frac{40{,}000 - (18/4)}{2(18/4)}\right] = 36.48$$

Recall that reduced variation will result in a higher S/N value; here reduced variation is equivalent to improved repeatability. The remaining data collected in the trials would be analyzed in the same fashion to find the S/N ratio for that trial. An ANOVA would be done on the eight S/N values to determine which control factors affected the average S/N value and which levels of those factors were more consistent. The intent is to see the measurement system respond in a consistent, linear fashion since the signal, or part to be measured, was structured linearly (three equally spaced values).

The first three S/N values demonstrate some interesting points. The first and second trials have exactly the same repeatability problem from operator to operator when measuring the three test pieces. However, in trial 2 the average reading of the 0.050 test piece (S_1) is low, the average reading of the 0.150 test piece (S_2) is on target, and the average reading of the 0.250 test piece (S_3) is high. Intuitively, it would seem as if trial 2 were a poorer gauge; however, it is really more sensitive to the differences in the signal parts and just needs to be rescaled to provide the proper reading. The respective values indicate trial 2 to be a better measuring system; a higher S/N value is obtained in trial 2. Trial 3 is better yet than trials 1 or 2. Even though not as sensitive as trial 2 (lower slope than trial 2), the repeatability around a point is better.

Small changes in S/N ratio are important. For each 3 db S/N value increases, the error variation is cut in half relative to the variation due to the linear portion of signal (or mean when using the original S/N formula).

Measurement system example 2. In an automatic transmission with electronic shift control logic, a device is needed to sense the output shaft speed, which is proportional to vehicle velocity. When a shift should occur is partly a function of vehicle speed. To sense speed, a transducer converts a varying magnetic field to a voltage with a frequency proportional to vehicle speed. The transducer is a measurement device to detect output shaft speed.

The transducer has a magnetic core around which a coil of conductive wire is wound. The tip of the magnetic core is held close to a rotating wheel, which is attached to the output shaft. The wheel is made of magnetically conductive (ferrous) material with notches or teeth on the perimeter, as shown in Fig. 7-12. As the notches move past the tip of the magnetic core, the magnetic field within the coil changes, which creates a voltage spike across the ends of the coil wire. The changing speed of the output shaft causes a change in the frequency of the voltage; consequently, the electronic "brain" can use this frequency signal as a measure of vehicle speed. The electronic circuit would like to have frequency be a function only of output shaft speed and have the voltage spike to be consistent over the range of speeds measured.

A designed experiment could be utilized to study such control factors as wire diameter, wire material, number of coil turns, coil diameter, coil length, magnetic core strength, wheel material, tooth shape,

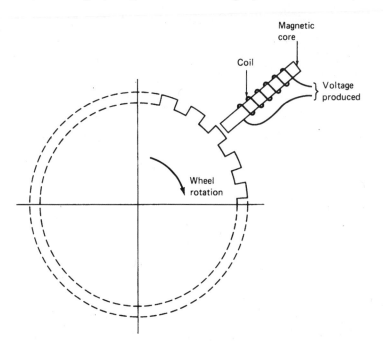

Figure 7-12 Speed sensor.

and tooth depth. Noise factors could be such things as core-to-wheel clearance (which may be difficult to hold closely in a transmission assembly), temperature, etc. The signal factor would be output shaft speed at three levels such as 500, 3500, and 6500 rpm. The peak voltage could be one response measured in the experiment and then the same type of S/N analysis and ANOVA carried out. This would identify which control factors provided consistent peak voltage at varying speeds, temperature, and clearances.

7-9-2   Heater control case study

The automobile heater control referenced in Sec. 7-9 could be evaluated using a parameter design approach with a dynamic characteristic for temperature control. A typical control mechanism might appear as in Fig. 7-13. There are several control factors possible with this design,

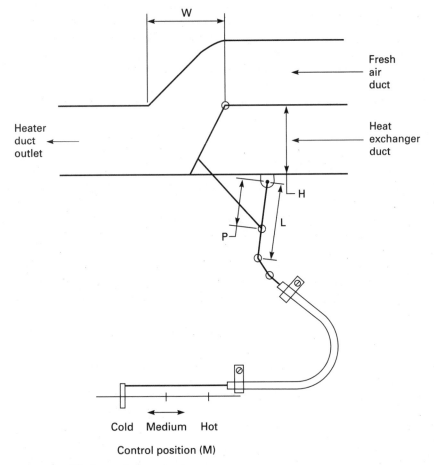

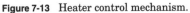

Control position (M)

Figure 7-13   Heater control mechanism.

but four of those factors were chosen to provide an example of parameter design using a dynamic characteristic. The dynamic characteristic is temperature of the air coming out of the vents. A linear relationship between the control position and the air temperature is preferred. The complete list of factors and levels involved with this experiment is:

Control factors	Level 1	Level 2	
W opening width	3.00 in	3.10	
H duct height	2.75 in	2.85	
L lever length	6.00 in	6.20	
P pivot length	3.00 in	3.10	

Signal factor	Level 1	Level 2	Level 3
M control position	1 (cold)	2 (mid)	3 (hot)

Noise factor	Level 1	Level 2
N outside temperature	30°F	60°F

The levels of the control factors are the proposed specification limits for the control system design.

A complete experimental layout and resultant data are shown in Table 7-20. Each trial of the OA has 12 total tests; each of the six

TABLE 7-20 Heater Control Experiment Summary

	Factors							Signal factor					
	W	H	L				P	M_1		M_2		M_3	
			Column no.							Noise factor			
Trial no.	1	2	3	4	5	6	7	N_1	N_2	N_1	N_2	N_1	N_2
1	1	1	1	1	1	1	1	31	69	89	111	103	127
1	1	1	1	1	1	1	1	26	64	86	110	104	132
2	1	1	1	2	2	2	2	39	59	75	111	107	132
2	1	1	1	2	2	2	2	24	60	74	114	94	137
3	1	2	2	1	1	2	2	35	61	69	117	96	121
3	1	2	2	1	1	2	2	24	56	74	114	97	132
4	1	2	2	2	2	1	1	39	51	80	116	102	137
4	1	2	2	2	2	1	1'	26	66	81	117	105	122
5	2	1	2	1	2	1	2	31	65	79	117	86	129
5	2	1	2	1	2	1	2	30	54	82	112	89	124
6	2	1	2	2	1	2	1	29	51	90	112	105	130
6	2	1	2	2	1	2	1	34	64	87	111	106	131
7	2	2	1	1	2	2	1	21	61	85	110	104	129
7	2	2	1	1	2	2	1	36	60	86	109	89	124
8	2	2	1	2	1	1	2	35	59	89	101	99	122
8	2	2	1	2	1	1	2	24	56	84	104	98	133

TABLE 7-21   Heater Control Experiment Trial 1

	Signal factor					
	M_1		M_2		M_3	
	Noise factor					
	N_1	N_2	N_1	N_2	N_1	N_2
Trial 1	31	69	89	111	103	127
	26	64	86	110	104	132
Totals	190		396		466	
Grand total			1052			

outer array factor combinations has two repetitions. The analysis follows the same format as the previous measurement system example 1. The first step is to perform an ANOVA on each set of trial results to determine the total sums of squares, sums of squares for linear component, and sums of squares for error (everything else). For the first trial, the data may be summarized as in Table 7-21 and the sums of squares calculated:

$$SS_T = 31^2 + 26^2 + \ldots + 132^2 - \left(\frac{1052^2}{12} \right) = 12{,}984.67$$

$$SS_{\text{linear}} = \frac{[-1(190) + 0(396) + 1(466)]^2}{2(4)} = 9522.00$$

$$SS_{\text{error}} = SS_T - SS_{\text{linear}} = 12{,}984.67 - 9522.00 = 3462.67$$

Variance for the linear and error components is

$$V_{\text{linear}} = \frac{SS_{\text{linear}}}{v_{\text{linear}}} = \frac{9522}{1} = 9522$$

$$V_{\text{error}} = \frac{SS_{\text{error}}}{v_{\text{error}}} = \frac{3462.67}{10} = 346.27$$

The S/N calculation is the same as the measurement system example:

$$S/N_1 = 10 \log \left[\frac{(V_{\text{linear}} - V_{\text{error}})}{(R \times V_{\text{error}})} \right]$$

$$S/N_1 = 10 \log \left[\frac{(9522.00 - 346.27)}{(4 \times 346.27)} \right] = 8.212$$

All eight S/N ratios are calculated in the same fashion and are summarized in Table 7-22. An ANOVA may then be performed using each

TABLE 7-22   Heater Control Experiment *S/N* Values

Trial no.	S/N
1	8.212
2	7.964
3	6.834
4	8.074
5	5.701
6	9.619
7	7.582
8	9.346

of the single S/N ratios in each trial as a quality characteristic. The unpooled and pooled ANOVA summaries are shown in Tables 7-23 and 7-24, respectively.

The factors and interactions that are statistically significant in order of decreasing effect are the lever length (L), width-lever length interaction $(W \times L)$, pivot length (P), and lever length-pivot length interaction $(L \times P)$. The preferred levels for lever length and pivot length are L_2 and P_1, respectively. The interaction combinations with

TABLE 7-23   Heater Control *S/N* ANOVA Summary Table

Source	SS	v	V	F	P
W	0.1682	1	0.1682	—	1.50
H	0.0144	1	0.0144	—	0.13
$W \times H / L \times P$	1.0368	1	1.0368	—	9.25
L	5.5445	1	5.5445	—	49.45
$W \times L / H \times P$	2.7378	1	2.7378	—	24.42
$H \times L / W \times P$	0.0544	1	0.0544	—	0.49
P	1.6562	1	1.6562	—	14.77
T	11.2123	7	—	—	100.00

TABLE 7-24   Heater Control Pooled *S/N* ANOVA Summary Table

Source	SS	v	V	F	P
$W \times H / L \times P$	1.0368	1	1.0368	13.1++	8.54
L	5.5445	1	5.5445	70.2#	48.74
$W \times L / H \times P$	2.7378	1	2.7378	34.6#	23.71
P	1.6562	1	1.6562	21.0++	14.07
e_p	0.2370	3	0.079	—	4.94
T	11.2123	7	—	—	100.00

+ at least 90% confidence
++ at least 95% confidence
# at least 99% confidence

the highest averages are W_2L_2 and L_2P_1. A plot of the vent air temperature versus control lever position (signal factor) would show a slightly nonlinear relationship for this design configuration.

Since the proposed specification limits were used for the levels of the control factors, the specification limits for lever length need to be increased, and for pivot length they need to be decreased, to obtain a more linear relationship. Both of these changes generally increase the S/N value, which means the linear component is becoming larger with respect to the error component. The same recommendation applies to the specification limits for opening width; the limits need to be increased to increase the S/N result as part of the interaction with lever length.

7-10   Summary

This chapter discussed some of the most powerful aspects of the Taguchi methodology, in particular, the parameter design concept. The strategy of making a product or process robust against noises by selecting the proper level for the appropriate parameters is the lowest-cost way of intentionally designing quality into a system. Various methods of designing an experiment with the noise effects taken into account were introduced. One method was to simply repeat the trials and let noise effects show up as they might, another approach utilized a noise array to force noise effects into the experiment, and yet another approach used the variation of the design parameters as a form of noise. A special case considering dynamic characteristics was covered in this chapter. This is probably the area of largest opportunity in the research and design activities of many new products. All of these approaches will lead to products and processes that are more robust against not only the noise included in the experiment, but against other noises as well.

The tolerance design approach of reducing the allowed parameter variation to reduce performance variation should be utilized only after the parameter design approach has proven to be insufficient.

Problems

7-1   Calculate the S/N_{LB} for this data from one inner array trial: 23, 25, 21, 20, 23, 19.

7-2   Calculate the S/N_{HB} for this data from one inner array trial: 110, 115, 112, 105.

7-3   Calculate the S/N_{NB_1} for this data from one inner array trial: 2.5, 2.9, 2.6, 2.7.

7-4 Calculate the S/N_{NB_2} for the same data as Prob. 7-3.

7-5 Calculate the log S value for the same data as Prob. 7-3.

7-6 What types of factors are usually class I, affecting both the average and variation of a quality characteristic?

7-7 What types of factors are usually class II, affecting only the variation of a quality characteristic?

7-8 What types of factors are usually class III, affecting only the average of a quality characteristic?

7-9 Calculate the S/N for this measurement system case:

Trial 1

M_1		M_2		M_3		(signal factor)
N_1	N_2	N_1	N_2	N_1	N_2	(noise factor)
0.101	0.102	0.500	0.502	0.900	0.901	
0.102	0.100	0.499	0.501	0.898	0.903	
0.100	0.103	0.503	0.502	0.901	0.904	

8

Tolerance Design

8-1 Introduction to Tolerance Design

Tolerance design is utilized when the efforts of parameter design have not proved adequate in reducing variation. In parameter design, low cost, widely varying components or factors may be used. If the quality of these components must still be improved to reduce variation to the desired amount, then tolerance design comes into play. Tolerance design requires the use of tighter tolerances or higher-quality, higher-cost materials to reduce variation to appropriate levels. In tolerance design, the loss function is used to substantiate the increased costs of higher-quality components by lower loss to society.

8-2 Tolerance Design Using the Loss Function

As an example, the transmission shift point discussion in Sec. 1-9 could be used to demonstrate the principle of tolerance design. The k value for the shift point adjustment was \$0.0625. The loss function for a nominal-is-best situation is

$$L(y) = k[S_y^2 + (y - m)^2]$$

In this case, it is assumed that the distribution of shift speeds will be centered because the variation around the average is the primary concern. Then the loss function simplifies to

$$L(y) = k \, S_y^2$$

If the current variance of shift speeds is 400 ($S_y = 20$ rpm), then the average loss per transmission is

$$L(y) = \$0.0625(400) = \$25.00$$

One of the components which contributes to this variation and subsequent loss is the shift valve spring. As a result of an experiment, the spring is known to have a substantial effect on shift speed, but parameter design approaches with other factors have not reduced the spring effect adequately. A higher-priced spring with less variance would reduce the shift variance to 200 ($S_y = 14$) at the additional cost of \$1.00 per spring (one per transmission). Would this be a wise investment? The new loss is

$$L(y) = \$0.0625(200) = \$12.50$$

Spending \$1.00 on the spring reduces the loss per transmission by \$12.50, according to the loss function, with a net gain of \$11.50 per transmission. Obviously, this is a substantial reduction in loss and is worthwhile.

The real trick in tolerance design is establishing the relationship of the variance of the component or factor to the variance of the performance characteristic of interest (i.e., spring variance affects shift variance by some relationship). If variance of the component or factor is reduced (quality improved) at some cost, then some reduction in variance of the performance characteristic will be obtained, resulting in a reduced loss, per the loss function.

The relationship between factor variances, factors A through F, and the total variance is described by the equation

$$S_T^2 = S_A^2 + S_B^2 + \dots + S_F^2 + S_e^2$$

The variance indicated here is the variance of the performance characteristic caused by the variance of the factor and not the actual variance of the factor itself. Recall how the variance of a factor is transmitted to the variance of the performance characteristic. The larger variances in this equation are caused by the more influential factors, and the lesser values of variance by the less influential factors.

The loss function uses the value of total variance in calculating an associated loss. If tolerance design is done on a less influential factor, then the total variance will not change that much and the loss function may not substantiate the tolerance design approach for that factor. The factors that contribute more toward the total variance are the ones that will be more effective in utilizing tolerance design.

8-3 Identification of Tolerance Design Factors

Two approaches may be utilized to identify which factors are candidates for the application of tolerance design (reduced variation of factor levels to reduce total variation). Both approaches are based on the use of two-level orthogonal arrays to determine the most important factors to control. The first approach uses standard deviation as a basis for the levels of factors to be evaluated. The second approach uses the specification limits as a basis for the levels of factors under investigation.

The factors that have an influence on a quality characteristic of concern are evaluated in an OA, as discussed in Chaps. 2 to 6. The analysis of variance is conducted as discussed in Chap. 5. The percent contribution is the key indicator of which factors are tolerance design candidates. Since meaningful measures of variation are used for all factors assigned to the OA, then the factors that cause the largest amount of variation must have tighter control to effectively reduce total variation.

Tighter limits of variation may be established and reevaluated to determine the adequacy of new limits to provide a desired amount of variation with respect to customer requirements. The loss function should be used to estimate the cost impacts of the tighter controls and the reduced variation that tighter limits provide.

Those factors that prove to be statistically insignificant or contribute a very small portion of the variation may have their limits loosened as a cost advantage.

For edification purposes, the Taguchi connotation of "tolerance" requires some explanation. Tolerance does not relate directly to the specification limits for a product or process. Taguchi uses tolerance in the sense of what variation of dimensions, components, and subassemblies will be allowed or tolerated by a system to perform as required. When thinking in terms of limits of variation, the specification limits do not directly reflect these limits. One product may have very good capability while another product may have poor capability with respect to a given dimension. The standard deviation or variance of a factor becomes a better measure of the actual variation present.

8-3-1 Setting levels for factors

The preferred method of determining candidate factors for tolerance design is based on standard deviation as levels of the factors in the experiment. Experiments using two or three factor levels may be conducted. The values for factor levels in those experiments are:

Number of levels	Factor level		
	1	2	3
Two	nominal $- 1\sigma$	nominal $+ 1\sigma$	
Three	nominal $- \sigma\sqrt{3/2}$	nominal	nominal $+ \sigma\sqrt{3/2}$

Factors with this amount of variation around the nominal provide an equivalent of 1 standard deviation using 2 or 3 data points. Since the independent parameters (factors, inputs) have variation equal to 1 standard deviation, then the amount of variation transmitted to the dependent parameters (responses, outputs, quality characteristics) will be directly proportional to the percent contribution of the total variation. This is a case where the total variance does have some informational value to the experimenter.

Product or process capability may be estimated using the total variance and improvements (reduced variation) of key factors may be evaluated using the total variance.

8-3-2   Tolerance design example

This case study is entirely hypothetical, but based on a mechanism with which most people are familiar: a lawn mower throttle control device. A throttle control mechanism is designed to be easily movable from a closed to an open throttle position. However, the throttle should remain in the chosen position until once again moved by the person using the mower. A certain amount of friction holds the throttle in the chosen position, but the friction cannot be too high or a person may not be able to move the lever. From a movement viewpoint, the friction should be considered to be a lower-is-better characteristic, but from a holding viewpoint, the friction should be thought of as a higher-is-better characteristic. This conflict of quality characteristics typically makes the actual operating window rather narrow.

A typical throttle control mechanism is shown in Fig. 8-1. The adjustment of the throttle position and, subsequently, the speed of the mower engine are accomplished by moving a lever attached to a cable which is linked to the carburetor throttle plate. By moving the control lever through a given angle, the cable is retracted or extended, which closes or opens the throttle plate of the carburetor. There are two main sources of friction which hold the control in a given position: the cable friction and the internal friction of the lever mechanism. The cable friction is not easily adjustable, so the lever mechanism is designed to provide a sufficient amount of friction to hold a specific position.

The lever mechanism is designed to create a certain amount of friction when two belleville springs are compressed, causing an interference in the completed assembly. This interference translates to a required force applied at the end of the lever to move the controls to a new position. By measuring the force at the end of the lever, the quality characteristic of interest to the final customer, the design may be evaluated for its effectiveness.

Tolerance design is used after parameter design has been applied.

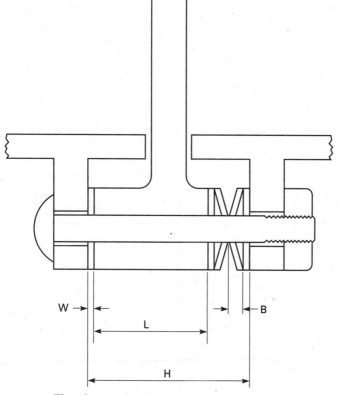

W →   ←   → ← B
L
H

Figure 8-1 Throttle control mechanism.

In this case, the mechanism would have been developed previously to select nominal parameter values that are in a less sensitive zone for lever force. However, let's assume some of the first models of the throttle mechanism showed problems in maintaining the chosen position, while others required uncomfortably high efforts to move the lever. A reduction in variation still would be necessary.

Initially, process capability studies should be conducted which might yield estimates of the standard deviation of various components of the throttle control mechanism as shown in Table 8-1. Using these estimates of standard deviation, the levels for these factors were set as shown in Table 8-2, plus and minus 1 standard deviation around the target nominal for a two-level experiment. All seven components could be assigned to a column in an L8 OA and the eight combinations evaluated twice to provide the results shown in Table 8-3. Lever force would be measured at the end of the lever where the person using the mower would apply effort to move the lever. The specification limits for lever force would be 2.0 to 3.0 lbs if a force greater

TABLE 8-1   Throttle Control Mechanism Dimensions

Components	Nominals, in	Standard deviation estimate, in
Washers (3)	0.040	0.005
Lever	0.750	0.010
Housing	1.000	0.010
Belleville springs (2)	0.150	0.005

TABLE 8-2   Throttle Control Experiment Factors and Levels

Components (factors)	Level 1, in	Level 2, in
W washers (3)	0.035	0.045
L lever	0.740	0.760
H housing	0.990	1.010
B belleville springs (2)	0.145	0.155

TABLE 8-3   Throttle Control Experiment Results

Trial no.	Lever force (lbs)	
	y_1	y_2
1	2.24	2.17
2	2.26	2.31
3	2.56	2.47
4	2.51	2.45
5	2.22	2.18
6	2.79	2.84
7	2.16	2.22
8	2.82	2.79

than 3 lbs would be uncomfortably high for many customers and a force less than 2 lbs would allow the lever to move due to mower vibrations.

An ANOVA of the raw data is summarized, unpooled and pooled, in Tables 8-4 and 8-5, respectively, and provides some key information about the throttle mechanism. First, there are two components (factors) that contribute large percentages to the total variation. The lever and the housing contribute over 75 percent of the observed variation. These are the two components that must have tighter control if variation is to be substantially reduced. Secondly, the total variance and standard deviation for lever force are calculated in this instance because this is the total variation produced by variation of all the significant components or factors. Total variance and standard deviation are calculated:

$$V_T = \frac{SS_T}{v_T} = \frac{0.975}{15} = 0.065$$

$$S_T = \sqrt{V_T} = \sqrt{0.065} = 0.25 \text{ lbs}$$

Six standard deviations, the product capability for lever force, is equal to

$$6S_T = 6(0.25) = 1.50 \text{ lbs}$$

This is greater than the width of the tolerance allowed for lever force; the upper specification is 3 lbs force and the lower specification is 2 lbs force.

To estimate the effect of reducing the variation of the tolerance design components or factors, an equation is written which comprehends the current situation. This equation is based on the current

TABLE 8-4   Throttle Lever Force ANOVA Summary

Source	SS	v	V	F	P
W_1	0.069	1	0.069	39.77#	6.89
W_2	0.059	1	0.059	33.93#	5.85
W_3	0.069	1	0.069	39.77#	6.89
L	0.406	1	0.406	234.54#	41.53
H	0.351	1	0.351	202.59#	35.85
B_1	0.003	1	0.003	1.91	0.16
B_2	0.003	1	0.003	1.91	0.16
e	0.015	8	0.002	—	2.67
T	0.975	15	0.065	—	100.00

+ at least 90% confidence
++ at least 95% confidence
# at least 99% confidence

TABLE 8-5   Throttle Lever Force Pooled ANOVA Summary

Source	SS	v	V	F	P
W_1	0.069	1	0.069	39.77#	6.86
W_2	0.059	1	0.059	33.93#	5.82
W_3	0.069	1	0.069	39.77#	6.86
L	0.406	1	0.406	234.54#	41.50
H	0.351	1	0.351	202.59#	35.81
e_p	0.021	10	0.002	—	3.15
T	0.975	15	0.065	—	100.00

+ at least 90% confidence
++ at least 95% confidence
# at least 99% confidence

total variance and the percent contribution of all the items in the ANOVA table, including error. Total variance is

$$V_T = V_T(P_{W_1} + P_{W_2} + P_{W_3} + P_L + P_H + P_{ep})$$
$$V_T = 0.065(0.0686 + 0.0582 + 0.0686 + 0.4150 + 0.3581 + 0.0315)$$
$$V_T = 0.065(1.00) = 0.065$$

Suppose the engineers determine that the standard deviation of the lever and housing can be reduced to half the present value by processing those components in a different fashion. The estimated total variance with tightened variation is then calculated:

$$V_T = 0.065[0.0686 + 0.0582 + 0.0686 + 1/4(0.4150) + 1/4(0.3581)$$
$$+ 0.0315]$$

$$V_T = 0.027$$

$$S_T = \sqrt{0.027} = 0.164 \text{ lbs}$$

$$6S_T = 6(0.164) = 0.984 \text{ lbs}$$

Reducing the variation of the two key components in the throttle mechanism is anticipated to make a substantial improvement in the total variation of lever force. The portion of variance due to the lever and housing is reduced by one-quarter since the standard deviation was reduced by one-half; variance is equal to standard deviation squared. The resultant estimated 6-standard-deviation spread of lever force is slightly smaller than the tolerance allowed for lever force.

In this case, a follow-up experiment could be run using reduced variation of the lever and housing. Table 8-6 shows a summary of the revised (tightened) variation for the throttle control mechanism and the new levels for components in an experiment. A new set of results was generated as shown in Table 8-7. An ANOVA of these results, summarized in Tables 8-8 and 8-9, shows that the lever and housing still contributed the largest percentages to total variation. Now, however, there is a much better balance of the contributions among the

TABLE 8-6   Throttle Control Experiment Factors and Levels

Components (factors)	Level 1, in	Level 2, in
W washers (3)	0.035	0.045
L lever*	0.745	0.755
H housing*	0.995	1.005
B belleville springs (2)	0.145	0.155

*Tighter controls are applied for improvement.

TABLE 8-7 Throttle Control Experiment Results

| Trial no. | Lever force (lbs) | |
	y_1	y_2
1	2.17	2.23
2	2.26	2.32
3	2.56	2.48
4	2.45	2.50
5	2.36	2.31
6	2.65	2.71
7	2.36	2.29
8	2.65	2.60

TABLE 8-8 Throttle Lever Force ANOVA Summary

Source	SS	v	V	F	P
W_1	0.058	1	0.058	31.10#	13.24
W_2	0.048	1	0.048	26.13#	11.05
W_3	0.081	1	0.081	43.86#	18.84
L	0.119	1	0.119	64.26#	27.82
H	0.090	1	0.090	48.64#	20.93
B_1	0.008	1	0.008	4.37+	1.48
B_2	0.002	1	0.002	1.10	0.04
e	0.015	8	0.002	—	6.60
T	0.421	15	0.028	—	100.00

+ at least 90% confidence
++ at least 95% confidence
# at least 99% confidence

TABLE 8-9 Throttle Lever Force Pooled ANOVA Summary

Source	SS	v	V	F	P
W_1	0.058	1	0.058	30.77#	13.23
W_2	0.048	1	0.048	25.85#	11.04
W_3	0.081	1	0.081	43.39#	18.84
L	0.119	1	0.119	63.59#	27.81
H	0.090	1	0.090	48.09#	20.93
B_1	0.008	1	0.008	4.32+	1.48
e_p	0.017	9	0.002	—	6.67
T	0.421	15	0.028	—	100.00

+ at least 90% confidence
++ at least 95% confidence
# at least 99% confidence

top five factors. Also, the standard deviation of the total variation of lever force is

$$S_T = \sqrt{\frac{0.421}{15}} = \sqrt{0.028} = 0.17 \text{ lbs}$$

The product capability for lever force, 6 standard deviations, is

$$6S_T = 6(0.17) = 1.02 \text{ lbs}$$

This is in agreement compared to the estimated standard deviation based on reduced component variation. Further tightening of component variation would be required in this case to consistently meet the stated specification limits for lever force.

This approach is based on the actual variation of known components or factors involved with the particular products or processes and is the preferred method of performing tolerance design. Using this approach, the actual capability can be directly estimated and validated.

8-3-3   Setting levels using specification limits

Another approach to tolerance design analysis is to use the specification limits (traditional tolerances) of the components or factors to determine the levels for the experiment. Using this approach, the key tolerance design candidates will be identified. This is equivalent to a sensitivity analysis or variation simulation to determine highly influential factors. In this case, knowledge of capability of individual components is not a requirement, but the total variation observed in the experiment will not necessarily be representative of real conditions. Therefore, actual capability cannot be estimated without evaluating the product or process after factor variation is reduced.

The levels for the components or factors will be set at the lower specification limit and the upper specification limit. This variation in component or factor values may not be realistic with respect to how the product or process typically runs; however, the analysis will still determine the relative sensitivity of the different components.

The experiment is then conducted and analyzed in the same manner as using the standard deviation–based levels. The assumption used in this approach is that the tolerance limits for the components or factors will be met so that the variation of quality characteristic of interest is meaningful. If the actual variation of some of the components of factors is greater than the tolerances, then wrong decisions will be made concerning which components need to be tightened for improved quality.

8-4 Quality Countermeasures

The earliest stages of design and development are the areas of greatest cost reduction in products and processes. Refer again to the three stages of design: system design (SD), parameter design (PD), and tolerance design (TD). Quality may be designed into a product or process by making it robust against all noises (outer, inner, and product), but only at certain stages of the product life cycle.

At the earliest stage of the life cycle, in research and development, SD may be used to improve quality with respect to all noises, as indicated by Table 8-10. One system relative to another may be more robust against noises. Once the system is chosen, then PD may be applied to combat noise effects also. TD can effectively be applied to inner and product noise, but PD should be used for outer noises.

When production engineering starts, the basic system has been selected and so have the nominal values for the design parameters. At this point, PD and TD are not very effective against outer and inner noises. Efforts from here through the actual production environment are effective against product noise.

Once the product is sold, no countermeasures are effective. Again, the loss function states that once a product has been shipped, there is some societal loss associated with that product. The bottom line is to start as early as possible in the product life cycle to make as high-quality, low-loss design as possible.

TABLE 8-10 Quality Countermeasures Against Noises

Area of quality control	Department	Countermeasure	Noises		
			Outer	Inner	Product
Off-line quality control	Research and development	System design	0	0	0
		Parameter design	0	0	0
		Tolerance design	0*	0	0
	Production engineering	System design	X	X	0
		Parameter design	X	X	0
		Tolerance design	X	X	0
On-line quality control	Manufacturing	Diagnosis and adjustment of processes	X	X	0
		Forecasting and correction	X	X	0
		Measurement and disposition	X	X	0
	Sales	Service	X	X	X

NOTE: 0 = effective; 0* = effective but not recommended; X = impossible.
SOURCE: Genichi Taguchi and Yu-in Wu, *Off-Line Quality Control,* Central Japan Quality Control Assoc., Nagaya, 1979, p. 80.

8-5 Summary

This chapter covers the final phase of the design process when tightened tolerances are a requirement to reduce variation to the desired amount. Tolerance design should be applied only after parameter design has not reduced variation sufficiently to meet customer expectations. The loss function should be applied to determine the cost effectiveness of reducing the variation of key components or factors.

Problems

8-1 What values would component levels assume in a two-level tolerance design experiment if the following nominals and standard deviations were available?

Component	Nominal	Standard deviation
A	1.250 in	0.002 in
B	1.254	0.002
C	1.500	0.003

8-2 What values would component levels assume in a three-level tolerance design experiment using the same information as Prob. 8-1?

8-3 Using the percent contribution and variance information in Table 8-5, what is the new total variance and standard deviation estimate if washer variation (std. dev.) is reduced by one-half?

8-4 Referring to Table 8-5 once more, what is the new total variance and standard deviation estimate if all components have variation reduced by one-half?

8-5 What comments may be made about the results obtained in Probs. 8-3 and 8-4 versus the results obtained in Sec. 8-3-2?

8-6 If the cost of replacing a throttle control mechanism to satisfy a customer is $30, what is the loss associated with the original design of the throttle mechanism, assuming the lever specification limits of 2.0 and 3.0 lbs force were meaningful to customers and the distribution was centered on the nominal value of 2.5 lbs?

8-7 What is the loss associated with the reduced variation produced by the tighter controls on the lever and housing in Sec. 8-3-2?

8-8 What comments may be made relative to the two loss values of Probs. 8-6 and 8-7?

Problem Solutions

Chapter 1

1-1

$$L = k(y - m)^2$$

$$k = \frac{L}{(y - m)^2} = \frac{\$0.02}{(1.3-1.2)^2} = 2.00$$

1-2

$$k = \frac{L}{(y - m)^2} = \frac{\$2.50}{(0.135-0.125)^2} = 25{,}000.0$$

1-3

$$L_1 = 25000\,(0.00333)^2 = \$0.28/\text{plate}$$
$$L_2 = 25000\,(0.00167)^2 = \$0.07/\text{plate}$$
$$L_3 = 25000\,[0.00167^2 + (0.120-0.125)^2] = \$0.69/\text{plate}$$

1-4

(a) Reduced variation decreases the associated loss per part.

(b) Missing the target is a serious loss compared to hitting the target with increased variation.

1-5

$$L_{LB} = k \cdot y^2$$

$$k = \frac{L}{y^2} = \frac{\$7.50}{0.002^2} = 1.875 \times 10^6$$

1-6

$$L_{HB} = k\left(\frac{1}{y^2}\right) = \frac{\$1.00}{(1/1000)^2} = 1 \times 10^6$$

1-7

$$L_{LB} = 1.875 \times 10^6 (0.0002^2 + 0.0014^2) = \$3.75/\text{part}$$

1-8

$$L_{HB} = 1 \times 10^6 \left(\frac{1}{1400^2}\right)\left[1 + \left(\frac{3 \times 30^2}{1400^2}\right)\right]$$

$$L_{HB} = \$0.51/\text{part}$$

1-9

$$(y - m)^2 = \frac{\$8.00}{0.0625} = 128$$

$$y - m = \pm 11 \text{ rpm}$$

Chapter 2

2-1

Planning, conducting, and analyzing

2-2

Stating problem

Stating objective

Selecting measurement system (quality characteristic)

Selecting factors

Selecting levels for factors

2-3

Positive information is the identification of factors and levels which do influence a quality characteristic. Negative information is the identification of factors which do *not* influence a quality characteristic, but no identification of factors which do influence a QC.

2-4

Factors are the independent parameters or inputs of a product or process which may be varied in a fixed or random fashion to cause a change in the dependent parameters, outputs, or quality characteristics.

2-5

Control factors are those factors that a manufacturer may set and which customers or environment do not directly change. Noise factors are those factors that a manufacturer cannot set or wishes not to set from a cost viewpoint.

2-6

Levels are the values in operational terms that are used to test or evaluate the factors' effect on outcome (e.g., two temperatures such as 250 and 300°F.

2-7

The purpose of flowcharts and C-E diagrams is to add some structure to a brainstorming session to document which factors may be investigated.

2-8

Variable characteristics are on a continuous scale, such as length, weight, speed, pressure, time, area, etc.

2-9

Attribute characteristics are on a discontinuous scale, being good or bad, such as nicks, scratches, dents, runs, cracks, smears, nits, etc.

Chapter 3

3-1

L8 and L16

3-2

Resolution 3

3-3

L16; this is the smallest array to accommodate eight factors with a resolution 2. The next larger array which provides resolution 3 is an L128, which is an eightfold increase in experimental size and probably not cost justified.

3-4

Resolution 4

3-5

Resolution 2

3-6

Column numbers 1, 2, 4, and 7

3-7

Column numbers 1, 2, 4, and 8

3-8

Two options with the same resolution are possible:

Column no.	1	2	3	4	5	6	7
Option 1	A	B	E	C	$A \times C$	$A \times D$	D
Option 2	A	C	$A \times C$	D	$A \times D$	B	E

3-9

L16; column numbers 1, 2, 4, 7, 8, 11, and 13

3-10

L16 and L32

3-11

Column numbers 1, 2, 5, and 9

3-12

Column numbers 1, 2, 3, 4, 5, and 6

3-13

These columns are groups of mutually interactive columns.

3-14

Column number 4

3-15

Column number 3; $A \times B = 5$, $C = 6$, $A \times B \times C = 5 \times 6 = 30$

3-16

Nesting

3-17

Column number 5 (interaction column of 1 and 4)

3-18

No, the combined factors approach converts 2 two-level factors to 1 three-level combined factor.

3-19

A smaller array may be used.

3-20

The product must be capable of reassembly and retest; the test must be nondestructive.

Chapter 4

4-1

The trial data sheet converts the OA symbology for levels (1, 2, 3, 4) to operational terms for the levels to prevent testing a wrong combination of levels, which would destroy the balance and orthogonality of the experiment.

4-2

Trial #7, water pump experiment

New cover design

New gasket design

LSL front bolt torque

No sealant

Smooth pump finish

USL back bolt torque

Front-to-back torque sequence

4-3

Trial #3, die-cast piston experiment

LSL copper content %

USL magnesium content %

LSL zinc content %

Die cooling off

Air cooling on

4-4

A much smaller sample size and total number of tests are required to detect statistically significant factors.

4-5

A 1.55 standard deviation shift

4-6

A 2.80 standard deviation shift

4-7

A 1.55 standard deviation shift

4-8

A total experimental sample size of 266, $N = 266$

4-9

A smaller sample size will be required to detect statistically significant factors.

4-10

To protect the experimental results from being biased by changes in unknown, uncontrolled factors which may influence the results

4-11

When it is expensive or difficult to change test conditions

4-12

When it is inexpensive or easy to change test conditions

4-13

That the influential factors have been included in the experiment and measurement error is small

4-14

That important factors have been excluded from the experiment or measurement error is large

Chapter 5

Practice Problem 1

$$SS_T = SS_m + SS_e$$

$$SS_T = 15^2 + 16^2 + \dots + 16^2 = 1822.0$$

$$SS_m = \frac{120^2}{8} = 1800.0$$

$$SS_e = 1822.0 - 1800.0 = 22.0$$

Practice Problem 2

$$SS_T = SS_A + SS_e$$

$$SS_T = 0.8^2 + 1.2^2 + \dots + 2.1^2 - \left(\frac{18.3^2}{12} \right) = 2.4825$$

$$SS_A = \left(\frac{4.2^2}{4} \right) + \left(\frac{5.9^2}{4} \right) + \left(\frac{8.2^2}{4} \right) - \left(\frac{18.3^2}{12} \right) = 2.0150$$

$$SS_e = 2.4825 - 2.0150 = 0.4675$$

5-1

Source	SS	v	V
m	622.29	1	622.29
e	17.71	6	2.95
T	640.00	7	—

5-2

Source	SS	v	V	F
m	282.69	3	94.23	58.17#
e	16.17	10	1.67	—
T	640.00	7	—	—

At least 99% confidence

5-3

Source	SS	v	V	F
$A*$	0.125	1	0.125	—
B	21.125	1	21.125	29.10#
$A \times B$	15.125	1	15.125	20.90#
$e*$	3.500	4	0.875	—
T	39.875	7	—	—
e_p*	3.625	5	0.725	—

At least 99% confidence

5-4

$$SS_A = \left(\frac{7^2}{2}\right) + \left(\frac{11^2}{2}\right) + \left(\frac{19^2}{2}\right) + \left(\frac{16^2}{2}\right) - \left(\frac{53^2}{8}\right)$$

$$SS_A = 42.375$$

5-5

$$SS_B = \left(\frac{7^2}{2}\right) + \left(\frac{3^2}{4}\right) + \left(\frac{13^2}{2}\right) - \left(\frac{23^2}{8}\right) = 45.125$$

$$SS_e = (1-2)^2/4 = 0.25$$

5-6

$$SS_{A_1} = \frac{[-3(9) - 1(3) + 1(12) + 3(19)]^2}{2(20)} = 38.025$$

$$SS_{A_q} = \frac{[+1(9) - 1(3) - 1(12) + 1(19)]^2}{2(4)} = 21.125$$

$$SS_{A_c} = \frac{[-1(9)+3(3)-3(12)+1(19)]^2}{2(20)} = 7.225$$

5-7

Source	SS	v	V	F
A	4.51	1	4.51	21.48#
B	0.01	1	0.01	—
C	0.11	1	0.11	—
D	0.11	1	0.11	—
E	0.11	1	0.11	—
F	0.01	1	0.01	—
G	0.11	1	0.11	—
e	14.92	72	0.21	—
T	19.89	79	—	—

At least 99% confidence

5-7

Source	SS	v	V	F
A	14.36	3	4.79	5.10#
B	10.68	3	3.56	3.79++
C	1.65	3	0.55	—
D	0.50	3	0.17	—
E	0.92	3	0.31	—
F	0.92	3	0.31	—
G	0.50	3	0.17	—
e	90.45	96	0.94	—
T	120.00	117	—	—

++ At least 95% confidence
# At least 99% confidence

Chapter 6

6-1

$$\hat{\mu}_{A_2B_1E_2F_2} = \overline{A}_2 + \overline{B}_1 + \overline{E}_2 + \overline{F}_2 - 3\overline{T}$$

6-2

$$\hat{\mu}_{A_2B_1E_2F_2} = 55 + 53 + 50 + 42 - 3(35) = 95$$

6-3

$$\overline{D_2E_2} = \frac{(72 + 72 + 72 + 71 + 69 + 71 + 76 + 74 + 74 + 72 + 74 + 74)}{12} = 72.58$$

6-4

$$F_{.10;1;12} = 3.18$$

6-5

$$CI_1 = \sqrt{\frac{(F\, V_e)}{n}} \qquad F_{.10;1;42} = 2.83$$

$$CI_1 = \sqrt{\frac{2.83(9.022)}{24}} \qquad V_e = 9.022$$

$$CI_1 = 1.03 \qquad \nu_e = 42$$

$$73.67 - 1.03 < \mu_{D_2} < 73.67 + 1.03 \qquad \overline{D}_2 = 73.67$$

$$72.64 < \mu_{D_2} < 74.70 \qquad n = 24$$

6-6

$$CI_2 = \sqrt{\frac{(F\, V_e)}{n_{\text{eff}}}} \qquad F_{.10;1;42} = 2.83$$

$$CI_2 = \sqrt{\frac{2.83(9.022)}{12}} \qquad V_e = 9.022$$

$$CI_2 = 1.46 \qquad n_{\text{eff}} = 12$$

$$78.16 - 1.46 < \mu_{A_2B_1D_2E_1} < 78.16 + 1.46$$

$$76.70 < \mu_{A_2B_1D_2E_1} < 79.62$$

Chapter 7

7-1

$$S/N_{\text{LB}} = -10 \log\left[\frac{1}{6\,(23^2 + 25^2 + 21^2 + 20^2 + 23^2 + 19^2)}\right]$$

$$S/N_{\text{LB}} = -26.82$$

7-2

$$S/N_{\text{HB}} = -10 \log\left\{\left[\frac{1}{4}\right]\left[\left(\frac{1}{110^2}\right) + \left(\frac{1}{115^2}\right) + \left(\frac{1}{112^2}\right) + \left(\frac{1}{105^2}\right)\right]\right\}$$

$$S/N_{\text{HB}} = 40.85$$

7-3

$$S/N_{\text{NB}_1} = -10 \log V_e = -10 \log (0.0292) = 15.35$$

7-4

$$S/N_{\text{NB}_2} = +10 \log\left[\frac{(SS_m - V_e)}{(r \times V_e)}\right]$$

$$S/N_{NB_2} = +10 \log \left[\frac{(28.6225 - 0.0292)}{(4 \times 0.0292)} \right]$$

$$S/N_{NB_2} = 23.89$$

7-5

$$\log S = -0.768$$

7-6

Factors that are nonlinear over an operational range

7-7

Factors that interact with noise factors

7-8

Factors that are linear over an operational range

7-9

$$S/N_{Trial\ 1} = +10 \log \left[\frac{(V_S - V_e)}{(R \times V_e)} \right]$$

$$= +10 \log \left[\frac{(1.9192 - 2.569 \times 10^{-6})}{(6 \times 2.569 \times 10^{-6})} \right]$$

$$S/N_{Trial\ 1} = 50.95$$

Chapter 8

8-1

	Level 1	Level 2
Component A	1.248	1.252
Component B	1.252	1.256
Component C	1.497	1.503

8-2

	Level 1	Level 2	Level 3
Component A	1.2476	1.250	1.2524
Component B	1.2516	1.254	1.2564
Component C	1.4963	1.500	1.5037

8-3

$$V_W = 0.065\left[\left(\frac{1}{4}\right)(0.0686 + 0.0582 + 0.0686) + 0.4150 + 0.3581 + 0.0315\right]$$

$$= 0.055$$

$$S_W = 0.236 \text{ lbs}$$

8-4

$$V_{all} = 0.065\left[\left(\frac{1}{4}\right)(0.0686 + 0.0582 + 0.0686 + 0.4150 + 0.3581) + 0.0315\right]$$
$$= 0.0178$$

$$S_{all} = 0.133 \text{ lbs}$$

8-5

$$S_{original} = 0.250 \text{ lbs}$$
$$S_{L\&H} = 0.164 \text{ lbs}$$
$$S_W = 0.236 \text{ lbs}$$
$$S_{all} = 0.133 \text{ lbs}$$

The greatest reduction with the fewest tightened tolerances was with the lever and housing, which are the key components. Tightening tolerances on the washers doesn't reduce variation much at all.

Two-Level Orthogonal Arrays

L4 Standard Array*

Trial no.	Column no.		
	1	2	3
1	1	1	1
2	1	2	2
3	2	1	2
4	2	2	1

*Two-level arrays from Genichi Taguchi and Yu-in Wu, *Off-Line Quality Control,* Central Japan Quality Control Association, Nagaya, 1979, pp. 103–107.

L8 Standard Array

Trial no.	Column no.						
	1	2	3	4	5	6	7
1	1	1	1	1	1	1	1
2	1	1	1	2	2	2	2
3	1	2	2	1	1	2	2
4	1	2	2	2	2	1	1
5	2	1	2	1	2	1	2
6	2	1	2	2	1	2	1
7	2	2	1	1	2	2	1
8	2	2	1	2	1	1	2

L8 Modified Array (one four-level factor)

	Standard column no.				
	123	4	5	6	7
	Modified column no.				
Trial no.	1	2	3	4	5
1	1	1	1	1	1
2	1	2	2	2	2
3	2	1	1	2	2
4	2	2	2	1	1
5	3	1	2	1	2
6	3	2	1	2	1
7	4	1	2	2	1
8	4	2	1	1	2

L12 Standard Array (no specific interaction columns available)*

	Column no.										
Trial no.	1	2	3	4	5	6	7	8	9	10	11
1	1	1	1	1	1	1	1	1	1	1	1
2	1	1	1	1	1	2	2	2	2	2	2
3	1	1	2	2	2	1	1	1	2	2	2
4	1	2	1	2	2	1	2	2	1	1	2
5	1	2	2	1	2	2	1	2	1	2	1
6	1	2	2	2	1	2	2	1	2	1	1
7	2	1	2	2	1	1	2	2	1	2	1
8	2	1	2	1	2	2	2	1	1	1	2
9	2	1	1	2	2	2	1	2	2	1	1
10	2	2	2	1	1	1	1	2	2	1	2
11	2	2	1	2	1	2	1	1	1	2	2
12	2	2	1	1	2	1	2	1	2	2	1

*Any column assignment provides a resolution 1 experiment.

L12 Modified Array (one four-level factor, no interactions)*

	Standard column no.								
	123	4	5	6	7	8	9	10	11
	Modified column no.								
Trial no.	1	2	3	4	5	6	7	8	9
1	1	1	1	1	1	1	1	1	1
2	1	1	1	2	2	2	2	2	2
3	1	2	2	1	1	1	2	2	2
4	2	2	2	1	2	2	1	1	2
5	2	1	2	2	1	2	1	2	1
6	2	2	1	2	2	1	2	1	1
7	3	2	1	1	2	2	1	2	1
8	3	1	2	2	2	1	1	1	2
9	3	2	2	2	1	2	2	1	1
10	4	1	1	1	1	2	2	1	2
11	4	2	1	2	1	1	1	2	2
12	4	1	2	1	2	1	2	2	1

*Any column assignment provides a resolution 1 experiment.

L16 Standard Array

	Column no.														
Trial no.	1	2	3	4	5	6	7	8	9	10	11	12	13	14	15
1	1	1	1	1	1	1	1	1	1	1	1	1	1	1	1
2	1	1	1	1	1	1	1	2	2	2	2	2	2	2	2
3	1	1	1	2	2	2	2	1	1	1	1	2	2	2	2
4	1	1	1	2	2	2	2	2	2	2	2	1	1	1	1
5	1	2	2	1	1	2	2	1	1	2	2	1	1	2	2
6	1	2	2	1	1	2	2	2	2	1	1	2	2	1	1
7	1	2	2	2	2	1	1	1	1	2	2	2	2	1	1
8	1	2	2	2	2	1	1	2	2	1	1	1	1	2	2
9	2	1	2	1	2	1	2	1	2	1	2	1	2	1	2
10	2	1	2	1	2	1	2	2	1	2	1	2	1	2	1
11	2	1	2	2	1	2	1	1	2	1	2	2	1	2	1
12	2	1	2	2	1	2	1	2	1	2	1	1	2	1	2
13	2	2	1	1	2	2	1	1	2	2	1	1	2	2	1
14	2	2	1	1	2	2	1	2	1	1	2	2	1	1	2
15	2	2	1	2	1	1	2	1	2	2	1	2	1	1	2
16	2	2	1	2	1	1	2	2	1	1	2	1	2	2	1

L16 Modified Array (one four-level factor)

	Standard column no.												
	123	4	5	6	7	8	9	10	11	12	13	14	15
	Modified column no.												
Trial no.	1	2	3	4	5	6	7	8	9	10	11	12	13
1	1	1	1	1	1	1	1	1	1	1	1	1	1
2	1	1	1	1	1	2	2	2	2	2	2	2	2
3	1	2	2	2	2	1	1	1	1	2	2	2	2
4	1	2	2	2	2	2	2	2	2	1	1	1	1
5	2	1	1	2	2	1	1	2	2	1	1	2	2
6	2	1	1	2	2	2	2	1	1	2	2	1	1
7	2	2	2	1	1	1	1	2	2	2	2	1	1
8	2	2	2	1	1	2	2	1	1	1	1	2	2
9	3	1	2	1	2	1	2	1	2	1	2	1	2
10	3	1	2	1	2	2	1	2	1	2	1	2	1
11	3	2	1	2	1	1	2	1	2	2	1	2	1
12	3	2	1	2	1	2	1	2	1	1	2	1	2
13	4	1	2	2	1	1	2	2	1	1	2	2	1
14	4	1	2	2	1	2	1	1	2	2	1	1	2
15	4	2	1	1	2	1	2	2	1	2	1	1	2
16	4	2	1	1	2	2	1	1	2	1	2	2	1

L16 Modified Array (two four-level factors)

	Standard column no.										
	123	4812	5	6	7	9	10	11	13	14	15
	Modified column no.										
Trial no.	1	2	3	4	5	6	7	8	9	10	11
1	1	1	1	1	1	1	1	1	1	1	1
2	1	2	1	1	1	2	2	2	2	2	2
3	1	3	2	2	2	1	1	1	2	2	2
4	1	4	2	2	2	2	2	2	1	1	1
5	2	1	1	2	2	1	2	2	1	2	2
6	2	2	1	2	2	2	1	1	2	1	1
7	2	3	2	1	1	1	2	2	2	1	1
8	2	4	2	1	1	2	1	1	1	2	2
9	3	1	2	1	2	2	1	2	2	1	2
10	3	2	2	1	2	1	2	1	1	2	1
11	3	3	1	2	1	2	1	2	1	2	1
12	3	4	1	2	1	1	2	1	2	1	2
13	4	1	2	2	1	2	2	1	2	2	1
14	4	2	2	2	1	1	1	2	1	1	2
15	4	3	1	1	2	2	2	1	1	1	2
16	4	4	1	1	2	1	1	2	2	2	1

L32 Standard Array

Trial no.	1	2	3	4	5	6	7	8	9	10	11	12	13	14	15
1	1	1	1	1	1	1	1	1	1	1	1	1	1	1	1
2	1	1	1	1	1	1	1	1	1	1	1	1	1	1	1
3	1	1	1	1	1	1	1	2	2	2	2	2	2	2	2
4	1	1	1	1	1	1	1	2	2	2	2	2	2	2	2
5	1	1	1	2	2	2	2	1	1	1	1	2	2	2	2
6	1	1	1	2	2	2	2	1	1	1	1	2	2	2	2
7	1	1	1	2	2	2	2	2	2	2	2	1	1	1	1
8	1	1	1	2	2	2	2	2	2	2	2	1	1	1	1
9	1	2	2	1	1	2	2	1	1	2	2	1	1	2	2
10	1	2	2	1	1	2	2	1	1	2	2	1	1	2	2
11	1	2	2	1	1	2	2	2	2	1	1	2	2	1	1
12	1	2	2	1	1	2	2	2	2	1	1	2	2	1	1
13	1	2	2	2	2	1	1	1	1	2	2	2	2	1	1
14	1	2	2	2	2	1	1	1	1	2	2	2	2	1	1
15	1	2	2	2	2	1	1	2	2	1	1	1	1	2	2
16	1	2	2	2	2	1	1	2	2	1	1	1	1	2	2
17	2	1	2	1	2	1	2	1	2	1	2	1	2	1	2
18	2	1	2	1	2	1	2	1	2	1	2	1	2	1	2
19	2	1	2	1	2	1	2	2	1	2	1	2	1	2	1
20	2	1	2	1	2	1	2	2	1	2	1	2	1	2	1
21	2	1	2	2	1	2	1	1	2	1	2	2	1	2	1
22	2	1	2	2	1	2	1	1	2	1	2	2	1	2	1
23	2	1	2	2	1	2	1	2	1	2	1	1	2	1	2
24	2	1	2	2	1	2	1	2	1	2	1	1	2	1	2
25	2	2	1	1	2	2	1	1	2	2	1	1	2	2	1
26	2	2	1	1	2	2	1	1	2	2	1	1	2	2	1
27	2	2	1	1	2	2	1	2	1	1	2	2	1	1	2
28	2	2	1	1	2	2	1	2	1	1	2	2	1	1	2
29	2	2	1	2	1	1	2	1	2	2	1	2	1	1	2
30	2	2	1	2	1	1	2	1	2	2	1	2	1	1	2
31	2	2	1	2	1	1	2	2	1	1	2	1	2	2	1
32	2	2	1	2	1	1	2	2	1	1	2	1	2	2	1

Column no.															
1	1	1	1	2	2	2	2	2	2	2	2	2	2	3	3
6	7	8	9	0	1	2	3	4	5	6	7	8	9	0	1
1	1	1	1	1	1	1	1	1	1	1	1	1	1	1	1
2	2	2	2	2	2	2	2	2	2	2	2	2	2	2	2
1	1	1	1	1	1	1	1	2	2	2	2	2	2	2	2
2	2	2	2	2	2	2	2	1	1	1	1	1	1	1	1
1	1	1	1	2	2	2	2	1	1	1	1	2	2	2	2
2	2	2	2	1	1	1	1	2	2	2	2	1	1	1	1
1	1	1	1	2	2	2	2	2	2	2	2	1	1	1	1
2	2	2	2	1	1	1	1	1	1	1	1	2	2	2	2
1	1	2	2	1	1	2	2	1	1	2	2	1	1	2	2
2	2	1	1	2	2	1	1	2	2	1	1	2	2	1	1
1	1	2	2	1	1	2	2	2	2	1	1	2	2	1	1
2	2	1	1	2	2	1	1	1	1	2	2	1	1	2	2
1	1	2	2	2	2	1	1	1	1	2	2	2	2	1	1
2	2	1	1	1	1	2	2	2	2	1	1	1	1	2	2
1	1	2	2	2	2	1	1	2	2	1	1	1	1	2	2
2	2	1	1	1	1	2	2	1	1	2	2	2	2	1	1
1	2	1	2	1	2	1	2	1	2	1	2	1	2	1	2
2	1	2	1	2	1	2	1	2	1	2	1	2	1	2	1
1	2	1	2	1	2	1	2	2	1	2	1	2	1	2	1
2	1	2	1	2	1	2	1	1	2	1	2	1	2	1	2
1	2	1	2	2	1	2	1	1	2	1	2	2	1	2	1
2	1	2	1	1	2	1	2	2	1	2	1	1	2	1	2
1	2	1	2	2	1	2	1	2	1	2	1	1	2	1	2
2	1	2	1	1	2	1	2	1	2	1	2	2	1	2	1
1	2	2	1	1	2	2	1	1	2	2	1	1	2	2	1
2	1	1	2	2	1	1	2	2	1	1	2	2	1	1	2
1	2	2	1	1	2	2	1	2	1	1	2	2	1	1	2
2	1	1	2	2	1	1	2	1	2	2	1	1	2	2	1
1	2	2	1	2	1	1	2	1	2	2	1	2	1	1	2
2	1	1	2	1	2	2	1	2	1	1	2	1	2	2	1
1	2	2	1	2	1	1	2	2	1	1	2	1	2	2	1
2	1	1	2	1	2	2	1	1	2	2	1	2	1	1	2

L32 Modified Array (one four-level factor)

					Standard column no.									
1														
2							1	1	1	1	1	1	1	
Trial no.	3	4	5	6	7	8	9	0	1	2	3	4	5	6

					Modified column no.									
									1	1	1	1	1	
Trial no.	1	2	3	4	5	6	7	8	9	0	1	2	3	4
1	1	1	1	1	1	1	1	1	1	1	1	1	1	1
2	1	1	1	1	1	1	1	1	1	1	1	1	1	2
3	1	1	1	1	1	2	2	2	2	2	2	2	2	1
4	1	1	1	1	1	2	2	2	2	2	2	2	2	2
5	1	2	2	2	2	1	1	1	1	2	2	2	2	1
6	1	2	2	2	2	1	1	1	1	2	2	2	2	2
7	1	2	2	2	2	2	2	2	2	1	1	1	1	1
8	1	2	2	2	2	2	2	2	2	1	1	1	1	2
9	2	1	1	2	2	1	1	2	2	1	1	2	2	1
10	2	1	1	2	2	1	1	2	2	1	1	2	2	2
11	2	1	1	2	2	2	2	1	1	2	2	1	1	1
12	2	1	1	2	2	2	2	1	1	2	2	1	1	2
13	2	2	2	1	1	1	1	2	2	2	2	1	1	1
14	2	2	2	1	1	1	1	2	2	2	2	1	1	2
15	2	2	2	1	1	2	2	1	1	1	1	2	2	1
16	2	2	2	1	1	2	2	1	1	1	1	2	2	2
17	3	1	2	1	2	1	2	1	2	1	2	1	2	1
18	3	1	2	1	2	1	2	1	2	1	2	1	2	2
19	3	1	2	1	2	2	1	2	1	2	1	2	1	1
20	3	1	2	1	2	2	1	2	1	2	1	2	1	2
21	3	2	1	2	1	1	2	1	2	2	1	2	1	1
22	3	2	1	2	1	1	2	1	2	2	1	2	1	2
23	3	2	1	2	1	2	1	2	1	1	2	1	2	1
24	3	2	1	2	1	2	1	2	1	1	2	1	2	2
25	4	1	2	2	1	1	2	2	1	1	2	2	1	1
26	4	1	2	2	1	1	2	2	1	1	2	2	1	2
27	4	1	2	2	1	2	1	1	2	2	1	1	2	1
28	4	1	2	2	1	2	1	1	2	2	1	1	2	2
29	4	2	1	1	2	1	2	2	1	2	1	1	2	1
30	4	2	1	1	2	1	2	2	1	2	1	1	2	2
31	4	2	1	1	2	2	1	1	2	1	2	2	1	1
32	4	2	1	1	2	2	1	1	2	1	2	2	1	2

						Standard column no.								
1	1	1	2	2	2	2	2	2	2	2	2	2	3	3
7	8	9	0	1	2	3	4	5	6	7	8	9	0	1
						Modified column no.								
1	1	1	1	1	2	2	2	2	2	2	2	2	2	2
5	6	7	8	9	0	1	2	3	4	5	6	7	8	9
1	1	1	1	1	1	1	1	1	1	1	1	1	1	1
2	2	2	2	2	2	2	2	2	2	2	2	2	2	2
1	1	1	1	1	1	1	2	2	2	2	2	2	2	2
2	2	2	2	2	2	2	1	1	1	1	1	1	1	1
1	1	1	2	2	2	2	1	1	1	1	2	2	2	2
2	2	2	1	1	1	1	2	2	2	2	1	1	1	1
1	1	1	2	2	2	2	2	2	2	2	1	1	1	1
2	2	2	1	1	1	1	1	1	1	1	2	2	2	2
1	2	2	1	1	2	2	1	1	2	2	1	1	2	2
2	1	1	2	2	1	1	2	2	1	1	2	2	1	1
1	2	2	1	1	2	2	2	2	1	1	2	2	1	1
2	1	1	2	2	1	1	1	1	2	2	1	1	2	2
1	2	2	2	2	1	1	1	1	2	2	2	2	1	1
2	1	1	1	1	2	2	2	2	1	1	1	1	2	2
1	2	2	2	2	1	1	2	2	1	1	1	1	2	2
2	1	1	1	1	2	2	1	1	2	2	2	2	1	1
2	1	2	1	2	1	2	1	2	1	2	1	2	1	2
1	2	1	2	1	2	1	2	1	2	1	2	1	2	1
2	1	2	1	2	1	2	2	1	2	1	2	1	2	1
1	2	1	2	1	2	1	1	2	1	2	1	2	1	2
2	1	2	2	1	2	1	1	2	1	2	2	1	2	1
1	2	1	1	2	1	2	2	1	2	1	1	2	1	2
2	1	2	2	1	2	1	2	1	2	1	1	2	1	2
1	2	1	1	2	1	2	1	2	1	2	2	1	2	1
2	2	1	1	2	2	1	1	2	2	1	1	2	2	1
1	1	2	2	1	1	2	2	1	1	2	2	1	1	2
2	2	1	1	2	2	1	2	1	1	2	2	1	1	2
1	1	2	2	1	1	2	1	2	2	1	1	2	2	1
2	2	1	2	1	1	2	1	2	2	1	2	1	1	2
1	1	2	1	2	2	1	2	1	1	2	1	2	2	1
2	2	1	2	1	1	2	2	1	1	2	1	2	2	1
1	1	2	1	2	2	1	1	2	2	1	2	1	1	2

L32 Modified Array (two four-level factors)

						Standard column no.						
1	4											
2	8						1	1	1	1	1	1
3	12	5	6	7	9	0	1	3	4	5	6	7

Trial no. (Standard column header)

						Modified column no.						
									1	1	1	1
1	2	3	4	5	6	7	8	9	0	1	2	3

Trial no.	1	2	3	4	5	6	7	8	9	0	1	2	3
1	1	1	1	1	1	1	1	1	1	1	1	1	1
2	1	1	1	1	1	1	1	1	1	1	1	2	2
3	1	2	1	1	1	2	2	2	2	2	2	1	1
4	1	2	1	1	1	2	2	2	2	2	2	2	2
5	1	3	2	2	2	1	1	1	2	2	2	1	1
6	1	3	2	2	2	1	1	1	2	2	2	2	2
7	1	4	2	2	2	2	2	2	1	1	1	1	1
8	1	4	2	2	2	2	2	2	1	1	1	2	2
9	2	1	1	2	2	1	2	2	1	2	2	1	1
10	2	1	1	2	2	1	2	2	1	2	2	2	2
11	2	2	1	2	2	2	1	1	2	1	1	1	1
12	2	2	1	2	2	2	1	1	2	1	1	2	2
13	2	3	2	1	1	1	2	2	2	1	1	1	1
14	2	3	2	1	1	1	2	2	2	1	1	2	2
15	2	4	2	1	1	2	1	1	1	2	2	1	1
16	2	4	2	1	1	2	1	1	1	2	2	2	2
17	3	1	2	1	2	2	1	2	2	1	2	1	2
18	3	1	2	1	2	2	1	2	2	1	2	2	1
19	3	2	2	1	2	1	2	1	1	2	1	1	2
20	3	2	2	1	2	1	2	1	1	2	1	2	1
21	3	3	1	2	1	2	1	2	1	2	1	1	2
22	3	3	1	2	1	2	1	2	1	2	1	2	1
23	3	4	1	2	1	1	2	1	2	1	2	1	2
24	3	4	1	2	1	1	2	1	2	1	2	2	1
25	4	1	2	2	1	2	2	1	2	2	1	1	2
26	4	1	2	2	1	2	2	1	2	2	1	2	1
27	4	2	2	2	1	1	1	2	1	1	2	1	2
28	4	2	2	2	1	1	1	2	1	1	2	2	1
29	4	3	1	1	2	2	2	1	1	1	2	1	2
30	4	3	1	1	2	2	2	1	1	1	2	2	1
31	4	4	1	1	2	1	1	2	2	2	1	1	2
32	4	4	1	1	2	1	1	2	2	2	1	2	1

					Standard column no.								
1	1	2	2	2	2	2	2	2	2	2	2	3	3
8	9	0	1	2	3	4	5	6	7	8	9	0	1

					Modified column no.								
1	1	1	1	1	1	2	2	2	2	2	2	2	2
4	5	6	7	8	9	0	1	2	3	4	5	6	7
1	1	1	1	1	1	1	1	1	1	1	1	1	1
2	2	2	2	2	2	2	2	2	2	2	2	2	2
1	1	1	1	1	1	2	2	2	2	2	2	2	2
2	2	2	2	2	2	1	1	1	1	1	1	1	1
1	1	2	2	2	2	1	1	1	1	2	2	2	2
2	2	1	1	1	1	2	2	2	2	1	1	1	1
1	1	2	2	2	2	2	2	2	2	1	1	1	1
2	2	1	1	1	1	1	1	1	1	2	2	2	2
2	2	1	1	2	2	1	1	2	2	1	1	2	2
1	1	2	2	1	1	2	2	1	1	2	2	1	1
2	2	1	1	2	2	2	2	1	1	2	2	1	1
1	1	2	2	1	1	1	1	2	2	1	1	2	2
1	1	1	1	2	2	2	2	1	1	1	1	2	2
2	2	2	2	1	1	2	2	1	1	1	1	2	2
1	1	1	1	2	2	1	1	2	2	2	2	1	1
1	2	1	2	1	2	1	2	1	2	1	2	1	2
2	1	2	1	2	1	2	1	2	1	2	1	2	1
1	2	1	2	1	2	2	1	2	1	2	1	2	1
2	1	2	1	2	1	1	2	1	2	1	2	1	2
1	2	2	1	2	1	1	2	1	2	2	1	2	1
2	1	1	2	1	2	2	1	2	1	1	2	1	2
1	2	2	1	2	1	2	1	2	1	1	2	1	2
2	1	1	2	1	2	1	2	1	2	2	1	2	1
2	1	1	2	2	1	1	2	2	1	1	2	2	1
1	2	2	1	1	2	2	1	1	2	2	1	1	2
2	1	1	2	2	1	2	1	1	2	2	1	1	2
1	2	2	1	1	2	1	2	2	1	1	2	2	1
2	1	2	1	1	2	1	2	2	1	2	1	1	2
1	2	1	2	2	1	2	1	1	2	1	2	2	1
2	1	2	1	1	2	2	1	1	2	1	2	2	1
1	2	1	2	2	1	1	2	2	1	2	1	1	2

Two-Level Interaction Table (doesn't apply to L12)

	Column no.													
Column no.	2	3	4	5	6	7	8	9	10	11	12	13	14	15
1	3	2	5	4	7	6	9	8	11	10	13	12	15	14
2	—	1	6	7	4	5	10	11	8	9	14	15	12	13
3	—	—	7	6	5	4	11	10	9	8	15	14	13	12
4	—	—	—	1	2	3	12	13	14	15	8	9	10	11
5	—	—	—	—	3	2	13	12	15	14	9	8	11	10
6	—	—	—	—	—	1	14	15	12	13	10	11	8	9
7	—	—	—	—	—	—	15	14	13	12	11	10	9	8
8	—	—	—	—	—	—	—	1	2	3	4	5	6	7
9	—	—	—	—	—	—	—	—	3	2	5	4	7	6
10	—	—	—	—	—	—	—	—	—	1	6	7	4	5
11	—	—	—	—	—	—	—	—	—	—	7	6	5	4
12	—	—	—	—	—	—	—	—	—	—	—	1	2	3
13	—	—	—	—	—	—	—	—	—	—	—	—	3	2
14	—	—	—	—	—	—	—	—	—	—	—	—	—	1
15	—	—	—	—	—	—	—	—	—	—	—	—	—	—
16	—	—	—	—	—	—	—	—	—	—	—	—	—	—
17	—	—	—	—	—	—	—	—	—	—	—	—	—	—
18	—	—	—	—	—	—	—	—	—	—	—	—	—	—
19	—	—	—	—	—	—	—	—	—	—	—	—	—	—
20	—	—	—	—	—	—	—	—	—	—	—	—	—	—
21	—	—	—	—	—	—	—	—	—	—	—	—	—	—
22	—	—	—	—	—	—	—	—	—	—	—	—	—	—
23	—	—	—	—	—	—	—	—	—	—	—	—	—	—
24	—	—	—	—	—	—	—	—	—	—	—	—	—	—
25	—	—	—	—	—	—	—	—	—	—	—	—	—	—
26	—	—	—	—	—	—	—	—	—	—	—	—	—	—
27	—	—	—	—	—	—	—	—	—	—	—	—	—	—
28	—	—	—	—	—	—	—	—	—	—	—	—	—	—
29	—	—	—	—	—	—	—	—	—	—	—	—	—	—
30	—	—	—	—	—	—	—	—	—	—	—	—	—	—

Two-Level Interaction Table (doesn't apply to L12) *(Continued)*

Column no.									Column no.							
	16	17	18	19	20	21	22	23	24	25	26	27	28	29	30	31
1	17	16	19	18	21	20	23	22	25	24	27	26	29	28	31	30
2	18	19	16	17	22	23	20	21	26	27	24	25	30	31	28	29
3	19	18	17	16	23	22	21	20	27	26	25	24	31	30	29	28
4	20	21	22	23	16	17	18	19	28	29	30	31	24	25	26	27
5	21	20	23	22	17	16	19	18	29	28	31	30	25	24	27	26
6	22	23	20	21	18	19	16	17	30	31	28	29	26	27	24	25
7	23	22	21	20	19	18	17	16	31	30	29	28	27	26	25	24
8	24	25	26	27	28	29	30	31	16	17	18	19	20	21	22	23
9	25	24	27	26	29	28	31	30	17	16	19	18	21	20	23	22
10	26	27	24	25	30	31	28	29	18	19	16	17	22	23	20	21
11	27	26	25	24	31	30	29	28	19	18	17	16	23	22	21	20
12	28	29	30	31	24	25	26	27	20	21	22	23	16	17	18	19
13	29	28	31	30	25	24	27	26	21	20	23	22	17	16	19	18
14	30	31	28	29	26	27	24	25	22	23	20	21	18	19	16	17
15	31	30	29	28	27	26	25	24	23	22	21	20	19	18	17	16
16	—	1	2	3	4	5	6	7	8	9	10	11	12	13	14	15
17	—	—	3	2	5	4	7	6	9	8	11	10	13	12	15	14
18	—	—	—	1	6	7	4	5	10	11	8	9	14	15	12	13
19	—	—	—	—	7	6	5	4	11	10	9	8	15	14	13	12
20	—	—	—	—	—	1	2	3	12	13	14	15	8	9	10	11
21	—	—	—	—	—	—	3	2	13	12	15	14	9	8	11	10
22	—	—	—	—	—	—	—	1	14	15	12	13	10	11	8	9
23	—	—	—	—	—	—	—	—	15	14	13	12	11	10	9	8
24	—	—	—	—	—	—	—	—	—	1	2	3	4	5	6	7
25	—	—	—	—	—	—	—	—	—	—	3	2	5	4	7	6
26	—	—	—	—	—	—	—	—	—	—	—	1	6	7	4	5
27	—	—	—	—	—	—	—	—	—	—	—	—	7	6	5	4
28	—	—	—	—	—	—	—	—	—	—	—	—	—	1	2	3
29	—	—	—	—	—	—	—	—	—	—	—	—	—	—	3	2
30	—	—	—	—	—	—	—	—	—	—	—	—	—	—	—	1

Three-Level Orthogonal Arrays

L9 Standard Array*

Trial no.	Column no.			
	1	2	3	4
1	1	1	1	1
2	1	2	2	2
3	1	3	3	3
4	2	1	2	3
5	2	2	3	1
6	2	3	1	2
7	3	1	3	2
8	3	2	1	3
9	3	3	2	1

*Three-level arrays from Genichi Taguchi and Yu-in Wu, *Off-Line Quality Control,* Central Japan Quality Control Association, Nagaya, 1979, pp. 108–110.

L18 Standard Array (no specific interaction columns available)*

	Column no.							
Trial no.	1	2	3	4	5	6	7	8
1	1	1	1	1	1	1	1	1
2	1	1	2	2	2	2	2	2
3	1	1	3	3	3	3	3	3
4	1	2	1	1	2	2	3	3
5	1	2	2	2	3	3	1	1
6	1	2	3	3	1	1	2	2
7	1	3	1	2	1	3	2	3
8	1	3	2	3	2	1	3	1
9	1	3	3	1	3	2	1	2
10	2	1	1	3	3	2	2	1
11	2	1	2	1	1	3	3	2
12	2	1	3	2	2	1	1	3
13	2	2	1	2	3	1	3	2
14	2	2	2	3	1	2	1	3
15	2	2	3	1	2	3	2	1
16	2	3	1	3	2	3	1	2
17	2	3	2	1	3	1	2	3
18	2	3	3	2	1	2	3	1

*Only the interaction between columns 1 and 2 may be calculated by using the layout below; other interactions are not in any specific column.

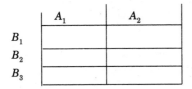

L27 Standard Array

Trial no.	Column no.												
	1	2	3	4	5	6	7	8	9	10	11	12	13
1	1	1	1	1	1	1	1	1	1	1	1	1	1
2	1	1	1	1	2	2	2	2	2	2	2	2	2
3	1	1	1	1	3	3	3	3	3	3	3	3	3
4	1	2	2	2	1	1	1	2	2	2	3	3	3
5	1	2	2	2	2	2	2	3	3	3	1	1	1
6	1	2	2	2	3	3	3	1	1	1	2	2	2
7	1	3	3	3	1	1	1	3	3	3	2	2	2
8	1	3	3	3	2	2	2	1	1	1	3	3	3
9	1	3	3	3	3	3	3	2	2	2	1	1	1
10	2	1	2	3	1	2	3	1	2	3	1	2	3
11	2	1	2	3	2	3	1	2	3	1	2	3	1
12	2	1	2	3	3	1	2	3	1	2	3	1	2
13	2	2	3	1	1	2	3	2	3	1	3	1	2
14	2	2	3	1	2	3	1	3	1	2	1	2	3
15	2	2	3	1	3	1	2	1	2	3	2	3	1
16	2	3	1	2	1	2	3	3	1	2	2	3	1
17	2	3	1	2	2	3	1	1	2	3	3	1	2
18	2	3	1	2	3	1	2	2	3	1	1	2	3
19	3	1	3	2	1	3	2	1	3	2	1	3	2
20	3	1	3	2	2	1	3	2	1	3	2	1	3
21	3	1	3	2	3	2	1	3	2	1	3	2	1
22	3	2	1	3	1	3	2	2	1	3	3	2	1
23	3	2	1	3	2	1	3	3	2	1	1	3	2
24	3	2	1	3	3	2	1	1	3	2	2	1	3
25	3	3	2	1	1	3	2	3	2	1	2	1	3
26	3	3	2	1	2	1	3	1	3	2	3	2	1
27	3	3	2	1	3	2	1	2	1	3	1	3	2

Three-Level Interaction Table (doesn't apply to L18)

Column nos.	Column no.											
	2	3	4	5	6	7	8	9	10	11	12	13
1	3	2	2	6	5	5	9	8	8	12	11	11
1	4	4	3	7	7	6	10	10	9	13	13	12
2	—	1	1	8	9	10	5	6	7	5	6	7
2	—	4	3	11	12	13	11	12	13	8	9	10
3	—	—	1	9	10	8	7	5	6	6	7	5
3	—	—	2	13	11	12	12	13	11	10	8	9
4	—	—	—	10	8	9	6	7	5	7	5	6
4	—	—	—	12	13	11	13	11	12	9	10	8
5	—	—	—	—	1	1	2	3	4	2	4	3
5	—	—	—	—	7	6	11	13	12	8	10	9
6	—	—	—	—	—	1	4	2	3	3	2	4
6	—	—	—	—	—	5	13	12	11	10	9	8
7	—	—	—	—	—	—	3	4	2	4	3	2
7	—	—	—	—	—	—	12	11	13	9	8	10
8	—	—	—	—	—	—	—	1	1	2	3	4
8	—	—	—	—	—	—	—	10	9	5	7	6
9	—	—	—	—	—	—	—	—	1	4	2	3
9	—	—	—	—	—	—	—	—	8	7	6	5
10	—	—	—	—	—	—	—	—	—	3	4	2
10	—	—	—	—	—	—	—	—	—	6	5	7
11	—	—	—	—	—	—	—	—	—	—	1	1
11	—	—	—	—	—	—	—	—	—	—	13	12
12	—	—	—	—	—	—	—	—	—	—	—	1
12	—	—	—	—	—	—	—	—	—	—	—	11

Design and Analysis Tables

TABLE D-1 Two-Level Orthogonal Array Selection

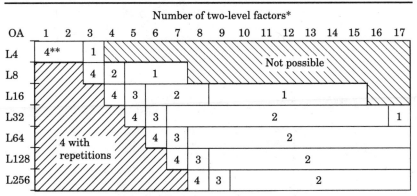

	Number of two-level factors*																
OA	1	2	3	4	5	6	7	8	9	10	11	12	13	14	15	16	17
L4	4**		1														
L8			4	2		1											
L16				4	3		2					1					
L32					4	3						2					1
L64						4	3						2				
L128							4	3						2			
L256								4	3					2			

*Consider each three- or four-level factor equivalent to 3 two-level factors; for example, 1 four-level factor and 5 two-level factors would have to use an L16 because that is equivalent to a total of 8 two-level factors

**Resolution number is a measure of the amount of confounding in a column:

4 = all items are in separate columns (full factorial)

3 = A and $B \times C \times D \times E$, or $A \times B$ and $C \times D \times E$ are in the same colum

2 = A and $B \times C \times D$, or $A \times B$ and $C \times D$ are in the same column

1 = A and $B \times C$ are in the same column

TABLE D-2   Three-Level Orthogonal Array Selection

Number of three-level factors

OA	1	2	3	4	5	6	7	8
L9	4*		1		Not possible			
L18				1				
L27	4			2		1		

*Resolution number is a measure of the amount of confounding in a column; see Table D-1.

TABLE D-3   Two-Level Orthogonal Array Factor Assignment

OA	Number of factors	Use column nos.	Resolution number*
L4	1–2	1,2	4 high
	3	1–3	1 low
L8	1–3	1,2,4	4 high
	4	1,2,4,7	2
	5–7	1,2,4,7,(3,5,6)†	1 low
L12	1–11	1–11	1 low
L16	1–4	1,2,4,8	4 high
	5	1,2,4,8,15	3
	6–8	1,2,4,7,8,(11,13,14)	2
	9–15	1,2,4,7,8,11,13,14,(3,5,6,9,10,12,15)	1 low
L32	1–5	1,2,4,8,16	4 high
	6	1,2,4,8,16,31	3
	7–16	1,2,4,8,16,31,(7,11,13,14,19,21,22, 25,26,28)	2
	17–31	1,2,4,7,8,11,13,14,16,19,21,22,25, 26,28,31,(3,5,6,9,10,12,15,17, 18,20,23,24,27,29,30)	1 low

*Resolution number is a measure of the amount of confounding in a column; see Table D-1.
†Column numbers in parentheses may be assigned in any order to achieve the indicated resolution; column numbers not in parentheses must be used first.

TABLE D-4   Three-Level Orthogonal Array Factor Assignment

OA	Number of factors	Use column nos.	Resolution number*
L9	1–2	1,2	4 high
	3–4	(1,2,3,4)†	1 low
L18	1–8	1–8	1 low
L27	1–3	1,2,5	4 high
	4	1,2,5,(9,10,12,13)	2
	5–13	1,2,3,4,5,(6–13)	1 low

*Resolution number is a measure of the amount of confounding in a column; see Table D-1.
†Column numbers in parentheses may be assigned in any order to achieve the indicated resolution; column numbers not in parentheses must be used first.

TABLE D-5   Polynomial Decomposition

Coefficient	$K = 2$	$K = 3$		$K = 4$			$K = 5$			
	L	L	Q	L	Q	C	L	Q	C	F
W_1	-1	-1	1	-3	1	-1	-2	2	-1	1
W_2	1	0	-2	-1	-1	3	-1	-1	2	-4
W_3		1	1	1	-1	-3	0	-2	0	6
W_4				3	1	1	1	-1	-2	-4
W_5							2	2	1	1
W_T	2	2	6	20	4	20	10	14	10	70

$$\text{SS}_{A \text{ polynomial}} = \frac{(W_1 \times A_1 + \ldots + W_K \times A_K)^2}{W_T \times R} = \text{Sum of squares}$$

NOTE: R = number of replications within a level (must be equal); L = linear; Q = quadratic; C = cubic; F = fourth.

TABLE D-6   *F* Values

$$F_{.10;\nu_1;\nu_2} \qquad 90\% \text{ confidence}^\dagger$$

		Degrees of freedom for the numerator (ν_1)								
	1	2	3	4	5	6	7	8	9	10
1	39.9	49.5	53.6	55.8	57.2	58.2	58.9	59.4	59.9	60.2
2	8.53	9.00	9.16	9.24	9.29	9.33	9.35	9.37	9.38	9.39
3	5.54	5.46	5.39	5.34	5.31	5.28	5.27	5.25	5.24	5.23
4	4.54	4.32	4.19	4.11	4.05	4.01	3.98	3.95	3.94	3.92
5	4.06	3.78	3.62	3.52	3.45	3.40	3.37	3.34	3.32	3.30
6	3.78	3.46	3.29	3.18	3.11	3.05	3.01	2.98	2.96	2.94
7	3.59	3.26	3.07	2.96	2.88	2.83	2.78	2.75	2.72	2.70
8	3.46	3.11	2.92	2.81	2.73	2.67	2.62	2.59	2.56	2.54
9	3.36	3.01	2.81	2.69	2.61	2.55	2.51	2.47	2.44	2.42
10	3.28	2.92	2.73	2.61	2.52	2.46	2.41	2.38	2.35	2.32
11	3.23	2.86	2.66	2.54	2.45	2.39	2.34	2.30	2.27	2.25
12	3.18	2.81	2.61	2.48	2.39	2.33	2.28	2.24	2.21	2.19
13	3.14	2.76	2.56	2.43	2.35	2.28	2.23	2.20	2.16	2.14
14	3.10	2.73	2.52	2.39	2.31	2.24	2.19	2.15	2.12	2.10
15	3.07	2.70	2.49	2.36	2.27	2.21	2.16	2.12	2.09	2.06
16	3.05	2.67	2.46	2.33	2.24	2.18	2.13	2.09	2.06	2.03
17	3.03	2.64	2.44	2.31	2.22	2.15	2.10	2.06	2.03	2.00
18	3.01	2.62	2.42	2.29	2.20	2.13	2.08	2.04	2.00	1.98
19	2.99	2.61	2.40	2.27	2.18	2.11	2.06	2.02	1.98	1.96
20	2.97	2.59	2.38	2.25	2.16	2.09	2.04	2.00	1.96	1.94
22	2.95	2.56	2.35	2.22	2.13	2.06	2.01	1.97	1.93	1.90
24	2.93	2.54	2.33	2.19	2.10	2.04	1.98	1.94	1.91	1.88
26	2.91	2.52	2.31	2.17	2.08	2.01	1.96	1.92	1.88	1.86
28	2.89	2.50	2.29	2.16	2.06	2.00	1.94	1.90	1.87	1.84
30	2.88	2.49	2.28	2.14	2.05	1.98	1.93	1.88	1.85	1.82
40	2.84	2.44	2.23	2.09	2.00	1.93	1.87	1.83	1.79	1.76
50	2.81	2.41	2.20	2.06	1.97	1.90	1.84	1.80	1.76	1.73
60	2.79	2.39	2.18	2.04	1.95	1.87	1.82	1.77	1.74	1.71
80	2.77	2.37	2.15	2.02	1.92	1.85	1.79	1.75	1.71	1.68
100	2.76	2.36	2.14	2.00	1.91	1.83	1.78	1.73	1.70	1.66
200	2.73	2.33	2.11	1.97	1.88	1.80	1.75	1.70	1.66	1.63
500	2.72	2.31	2.10	1.96	1.86	1.79	1.73	1.68	1.64	1.61
∞	2.71	2.30	2.08	1.94	1.85	1.77	1.72	1.67	1.63	1.60

Degrees of freedom for the denominator (ν_2)

$$F_{.05;\nu_1;\nu_2} \qquad 95\% \text{ confidence}^\ddagger$$

		Degrees of freedom for the numerator (ν_1)								
	1	2	3	4	5	6	7	8	9	10
1	161	200	216	225	230	234	237	239	241	242
2	18.5	19.0	19.2	19.2	19.3	19.3	19.4	19.4	19.4	19.4
3	10.1	9.55	9.28	9.12	9.01	8.94	8.89	8.85	8.81	8.79
4	7.71	6.94	6.59	6.39	6.26	6.16	6.09	6.04	6.00	5.96
5	6.61	5.79	5.41	5.19	5.05	4.95	4.88	4.82	4.77	4.74
6	5.99	5.14	4.76	4.53	4.39	4.28	4.21	4.15	4.10	4.06
7	5.59	4.74	4.35	4.12	3.97	3.87	3.79	3.73	3.68	3.64
8	5.32	4.46	4.07	3.84	3.69	3.58	3.50	3.44	3.39	3.35
9	5.12	4.26	3.86	3.63	3.48	3.37	3.29	3.23	3.18	3.14
10	4.96	4.10	3.71	3.48	3.33	3.22	3.14	3.07	3.02	2.98
11	4.84	2.98	3.50	3.36	3.20	3.01	2.95	2.90	2.85	2.82

Degrees of freedom for the denominator (ν_2)

TABLE D-6 *F* Values (Continued)

$$F_{.05;\nu_1;\nu_2} \qquad 95\% \text{ confidence}^{\dagger\dagger}$$

	Degrees of freedom for the numerator (ν_1)									
	1	2	3	4	5	6	7	8	9	10
12	4.75	3.89	3.49	3.26	3.11	3.00	2.91	2.85	2.80	2.75
13	4.67	3.81	3.41	3.18	3.03	2.92	2.83	2.77	2.71	2.67
14	4.60	3.74	3.34	3.11	2.96	2.85	2.76	2.70	2.65	2.60
15	4.54	3.68	3.29	3.06	2.90	2.79	2.71	2.64	2.59	2.54
16	4.49	3.63	3.24	3.01	2.85	2.74	2.66	2.59	2.54	2.49
17	4.45	3.59	3.20	2.96	2.81	2.70	2.61	2.55	2.49	2.45
18	4.41	3.55	3.16	2.93	2.77	2.66	2.58	2.51	2.46	2.41
19	4.38	3.52	3.13	2.90	2.74	2.63	2.54	2.48	2.42	2.38
20	4.35	3.49	3.10	2.87	2.71	2.60	2.51	2.45	2.39	2.35
21	4.32	3.47	3.07	2.82	2.68	2.57	2.49	2.42	2.37	2.32
22	4.30	3.44	3.05	2.84	2.66	2.55	2.46	2.40	2.34	2.30
23	4.28	3.42	3.03	2.80	2.64	2.53	2.44	2.37	2.32	2.27
24	4.26	3.40	3.01	2.78	2.62	2.51	2.42	2.36	2.30	2.25
25	4.24	3.39	2.99	2.76	2.60	2.49	2.40	2.34	2.28	2.24
26	4.23	3.37	2.98	2.74	2.59	2.47	2.39	2.32	2.27	2.22
27	4.21	3.35	2.96	2.73	2.57	2.46	2.37	2.31	2.25	2.20
28	4.20	3.34	2.95	2.71	2.56	2.45	2.36	2.29	2.24	2.19
29	4.18	3.33	2.93	2.70	2.55	2.43	2.35	2.28	2.22	2.18
30	4.17	3.32	2.92	2.69	2.53	2.42	2.33	2.27	2.21	2.16
32	4.15	3.29	2.90	2.67	2.51	2.40	2.31	2.24	2.19	2.14
34	4.13	3.28	2.88	2.65	2.49	2.38	2.29	2.23	2.17	2.12
36	4.11	3.26	2.87	2.63	2.48	2.36	2.28	2.21	2.15	2.11
38	4.10	3.24	2.85	2.62	2.46	2.35	2.26	2.19	2.14	2.09
40	4.08	3.23	2.84	2.61	2.45	2.34	2.25	2.18	2.12	2.08
42	4.07	3.22	2.83	2.59	2.44	2.32	2.24	2.16	2.11	2.06
44	4.06	3.21	2.82	2.58	2.43	2.31	2.23	2.16	2.10	2.05
46	4.05	3.20	2.81	2.57	2.42	2.30	2.22	2.15	2.09	2.04
48	4.04	3.19	2.80	2.57	2.41	2.29	2.21	2.14	2.08	2.03
50	4.03	3.18	2.79	2.56	2.40	2.29	2.20	2.13	2.07	2.03
55	4.02	3.16	2.77	2.54	2.38	2.27	2.18	2.11	2.06	2.01
60	4.00	3.15	2.76	2.53	2.37	2.25	2.17	2.10	2.04	1.99
65	3.99	3.14	2.75	2.51	2.36	2.24	2.15	2.08	2.03	1.98
70	3.98	3.13	2.74	2.50	2.35	2.23	2.14	2.07	2.02	1.97
80	3.96	3.11	2.73	2.49	2.33	2.21	2.13	2.06	2.00	1.95
90	3.95	3.10	2.71	2.47	2.32	2.20	2.11	2.0 4	1.99	1.94
100	3.94	3.09	2.70	2.46	2.31	2.19	2.10	2.03	1.97	1.93
125	3.92	3.07	2.68	2.44	2.29	2.17	2.08	2.01	1.96	1.91
150	3.90	3.08	2.66	2.43	2.27	2.16	2.07	2.00	1.94	1.89
200	3.89	3.04	2.65	2.42	2.26	2.14	2.06	1.98	1.93	1.88
300	3.87	3.03	2.63	2.40	2.24	2.13	2.04	1.97	1.91	1.86
500	3.86	3.01	2.62	2.39	2.23	2.12	2.03	1.96	1.90	1.85
1000	3.85	3.00	2.61	2.38	2.22	2.11	2.02	1.95	1.89	1.84
∞	3.84	3.00	2.60	2.37	2.21	2.10	2.01	1.94	1.88	1.83

Degrees of freedom for the denominator (ν_2)

TABLE D-6 *F* Values (Continued)

$$F_{.01;\nu_1;\nu_2} \quad 99\% \text{ confidence\#}$$

		Degrees of freedom for the numerator (ν_1)								
	1	2	3	4	5	6	7	8	9	10
	Multiply the numbers of the first row ($\nu_2 = 1$) by 10									
1	405	500	540	563	576	596	598	598	602	606
2	93.5	99.0	99.3	99.3	99.3	99.3	99.4	99.4	99.4	99.4
3	34.1	30.8	20.5	28.7	28.2	27.9	27.7	27.5	27.3	27.2
4	21.2	18.0	16.7	16.0	15.5	15.2	15.0	14.8	14.7	14.5
5	16.8	13.2	12.1	11.4	11.0	10.7	10.5	10.3	10.2	10.1
6	13.7	10.9	9.78	9.15	8.75	8.47	8.28	8.10	7.98	7.87
7	12.2	9.55	8.45	7.85	7.46	7.19	6.99	6.94	6.72	6.62
8	11.3	8.65	7.89	7.01	6.63	6.37	6.18	6.03	5.91	5.81
9	10.6	8.02	6.99	6.42	6.06	5.80	5.61	5.47	5.35	5.26
10	10.0	7.56	6.55	5.99	5.64	5.39	5.20	5.06	4.94	4.85
11	9.65	7.21	6.22	5.67	5.32	5.07	4.89	4.74	4.63	4.54
12	9.33	6.93	5.95	5.41	5.06	4.82	4.64	4.50	4.30	4.30
13	9.07	6.70	5.74	5.21	4.86	4.62	4.44	4.30	4.19	4.10
14	8.86	6.51	5.58	5.04	4.70	4.46	4.28	4.14	4.03	3.94
15	8.68	6.26	5.42	4.89	4.56	4.32	4.14	4.00	3.89	3.80
16	8.53	6.22	5.29	4.77	4.44	4.20	4.03	3.89	3.78	3.69
17	8.60	6.11	5.18	4.67	4.34	4.10	3.93	3.79	3.68	3.59
18	8.20	6.01	5.09	4.58	4.25	4.01	3.84	3.71	3.60	3.51
19	8.18	5.93	5.01	4.50	4.17	3.94	3.77	3.68	3.52	3.43
20	8.10	5.85	4.94	4.43	4.10	3.87	3.70	3.56	3.46	3.37
21	8.02	5.78	4.87	4.37	4.04	3.81	3.64	3.51	3.40	3.31
22	7.95	5.72	4.82	4.31	3.99	3.76	3.59	3.45	3.35	3.26
23	7.86	5.66	4.76	4.26	3.94	3.71	3.54	3.41	3.30	3.21
24	7.82	5.61	4.72	4.22	3.90	3.67	3.50	3.36	3.26	3.17
25	7.77	5.57	4.68	4.18	3.86	3.63	3.46	3.32	3.22	3.13
26	7.72	5.53	4.64	4.14	3.82	3.59	3.42	3.29	3.18	3.09
27	7.66	5.49	4.60	4.11	3.78	3.56	3.39	3.26	3.15	3.06
28	7.64	5.45	4.57	4.07	3.75	3.53	3.36	3.23	3.12	3.03
29	7.60	5.42	4.54	4.04	3.73	3.50	3.33	3.20	3.09	3.00
30	7.56	5.39	4.51	4.03	3.70	3.47	3.30	3.17	3.07	2.98
32	7.50	5.34	4.46	3.97	3.65	3.43	3.26	3.13	3.02	2.93
34	7.44	5.29	4.42	3.93	3.61	3.39	3.23	3.09	2.96	2.89
36	7.40	5.25	4.36	3.89	3.57	3.35	3.18	3.05	2.95	2.86
38	7.35	5.21	4.34	3.86	3.54	3.32	3.15	3.02	2.92	2.83
40	7.31	5.18	4.31	3.83	3.51	3.29	3.12	2.99	2.80	2.80
42	7.28	5.15	4.29	3.80	3.49	3.27	3.10	2.97	2.86	2.78
44	7.25	5.12	4.26	3.78	3.47	3.24	3.08	2.95	2.84	2.75
46	7.22	5.10	4.24	3.76	3.44	3.22	3.06	2.93	2.82	2.73
48	7.19	5.08	4.22	3.74	3.43	3.20	3.04	2.91	2.80	2.72
50	7.17	5.06	4.20	3.72	3.41	3.19	3.02	2.89	2.79	2.70
55	7.12	5.01	4.16	3.68	3.37	3.15	2.98	2.85	2.75	2.66
60	7.08	4.98	4.12	3.65	3.34	3.12	2.95	2.82	2.72	2.63
65	7.04	4.95	4.10	3.62	3.31	3.09	2.93	2.80	2.69	2.61
70	7.01	4.92	4.08	3.60	3.29	3.07	2.91	2.78	2.67	2.59
80	6.98	4.88	4.04	3.56	3.26	3.04	2.87	2.74	2.64	2.55
90	6.93	4.85	4.01	3.54	3.23	3.01	2.84	2.72	2.61	2.52
100	6.90	4.83	3.96	3.51	3.21	2.99	2.82	2.69	2.59	2.50

Degrees of freedom for the denominator (ν_2)

TABLE D-6 *F* Values (Continued)

$$F_{.01;\nu_1;\nu_2} \qquad 99\% \text{ confidence}^{\#}$$

Degrees of freedom for the denominator (ν_2)	Degrees of freedom for the numerator (ν_1)									
	1	2	3	4	5	6	7	8	9	10
125	6.84	4.78	3.94	3.47	3.17	2.95	2.79	2.66	2.55	2.47
150	6.81	4.75	3.92	3.45	3.14	2.92	2.76	2.63	2.53	2.44
200	6.76	4.71	3.88	3.41	3.11	2.89	2.73	2.60	2.50	2.41
300	6.72	4.68	3.85	3.38	3.08	2.86	2.70	2.57	2.47	2.36
500	6.69	4.65	3.82	3.36	3.05	2.84	2.68	2.55	2.44	2.36
1000	6.66	4.63	3.80	3.34	3.04	2.82	2.66	2.53	2.43	2.34
∞	6.66	4.61	3.78	3.32	3.02	2.80	2.64	2.51	2.41	2.32

TABLE D-7   Omega Conversion Table*

p, %	db	p, %	db	p,%	db
0.0	∞	5.0	−12.787	10.0	−9.541
0.1	−29.995	5.1	−12.696	10.1	−9.493
0.2	−26.980	5.2	−12.607	10.2	−9.446
0.3	−25.215	5.3	−12.520	10.3	−9.399
0.4	−23.961	5.4	−12.434	10.4	−9.352
0.5	−22.988	5.5	−12.350	10.5	−9.305
0.6	−22.191	5.6	−12.267	10.6	−9.259
0.7	−21.518	5.7	−12.185	10.7	−9.214
0.8	−20.933	5.8	−12.105	10.8	−9.168
0.9	−20.417	5.9	−12.026	10.9	−9.124
1.0	−19.955	6.0	−11.949	11.0	−9.079
1.1	−19.537	6.1	−11.872	11.1	−9.035
1.2	−19.155	6.2	−11.797	11.2	−8.991
1.3	−18.803	6.3	−11.723	11.3	−8.947
1.4	−18.476	6.4	−11.650	11.4	−8.904
1.5	−18.172	6.5	−11.578	11.5	−8.861
1.6	−17.888	6.6	−11.507	11.6	−8.819
1.7	−17.620	6.7	−11.437	11.7	−8.777
1.8	−17.367	6.8	−11.368	11.8	−8.735
1.9	−17.128	6.9	−11.300	11.9	−8.693
2.0	−16.901	7.0	−11.233	12.0	−8.652
2.1	−16.685	7.1	−11.167	12.1	−8.611
2.2	−16.478	7.2	−11.101	12.2	−8.570
2.3	−16.281	7.3	−11.037	12.3	−8.530
2.4	−16.091	7.4	−10.973	12.4	−8.490
2.5	−15.910	7.5	−10.910	12.5	−8.450
2.6	−15.735	7.6	−10.848	12.6	−8.410
2.7	−15.566	7.7	−10.786	12.7	−8.371
2.8	−15.404	7.8	−10.725	12.8	−8.332
2.9	−15.247	7.9	−10.665	12.9	−8.293
3.0	−15.096	8.0	−10.606	13.0	−8.255
3.1	−14.949	8.1	−10.547	13.1	−8.216
3.2	−14.806	8.2	−10.489	13.2	−8.178
3.3	−14.668	8.3	−10.432	13.3	−8.141
3.4	−14.534	8.4	−10.375	13.4	−8.103
3.5	−14.404	8.5	−10.319	13.5	−8.066
3.6	−14.227	8.6	−10.263	13.6	−8.029
3.7	−14.153	8.7	−10.209	13.7	−7.992
3.8	−14.033	8.8	−10.154	13.8	−7.955
3.9	−13.916	8.9	−10.100	13.9	−7.919
4.0	−13.801	9.0	−10.047	14.0	−7.883
4.1	−13.689	9.1	−9.994	14.1	−7.847
4.2	−13.580	9.2	−9.942	14.2	−7.811
4.3	−13.473	9.3	−9.890	14.3	−7.775
4.4	−13.369	9.4	−9.839	14.4	−7.740
4.5	−13.267	9.5	−9.788	14.5	−7.705
4.6	−13.167	9.6	−9.738	14.6	−7.670
4.7	−13.069	9.7	−9.688	14.7	−7.635
4.8	−12.973	9.8	−9.639	14.8	−7.601
4.9	−12.879	9.9	−9.590	14.9	−7.566

*Genichi Taguchi and Yu-In Wu, *Off-Line Quality Control,* Central Japan Quality Control Association, Nagaya, 1979, pp. 99–102.

TABLE D-7   Omega Conversion Table (*Continued*)

p, %	db	p, %	db	p,%	db
15.0	−7.532	20.0	−6.020	25.0	−4.770
15.1	−7.498	20.1	−5.993	25.1	−4.747
15.2	−7.465	20.2	−5.966	25.2	−4.724
15.3	−7.431	20.3	−5.939	25.3	−4.701
15.4	−7.397	20.4	−5.912	25.4	−4.678
15.5	−7.364	20.5	−5.885	25.5	−4.655
15.6	−7.331	20.6	−5.859	25.6	−4.632
15.7	−7.298	20.7	−5.832	25.7	−4.610
15.8	−7.266	20.8	−5.806	25.8	−4.587
15.9	−7.233	20.9	−5.779	25.9	−4.564
16.0	−7.201	21.0	−5.753	26.0	−4.542
16.1	−7.168	21.1	−5.727	26.1	−4.519
16.2	−7.136	21.2	−5.701	26.2	−4.497
16.3	−7.104	21.3	−5.675	26.3	−4.474
16.4	−7.073	21.4	−5.649	26.4	−4.452
16.5	−7.041	21.5	−5.623	26.5	−4.429
16.6	−7.010	21.6	−5.598	26.6	−4.407
16.7	−6.978	21.7	−5.572	26.7	−4.385
16.8	−6.947	21.8	−5.547	26.8	−4.363
16.9	−6.916	21.9	−5.521	26.9	−4.341
17.0	−6.885	22.0	−5.496	27.0	−4.319
17.1	−6.855	22.1	−5.470	27.1	−4.297
17.2	−6.824	22.2	−5.445	27.2	−4.275
17.3	−6.794	22.3	−5.420	27.3	−4.253
17.4	−6.763	22.4	−5.395	27.4	−4.231
17.5	−6.733	22.5	−5.370	27.5	−4.209
17.6	−6.703	22.6	−5.345	27.6	−4.187
17.7	−6.673	22.7	−5.321	27.7	−4.166
17.8	−6.644	22.8	−5.296	27.8	−4.144
17.9	−6.614	22.9	−5.271	27.9	−4.122
18.0	−6.584	23.0	−5.427	28.0	−4.101
18.1	−6.555	23.1	−5.222	28.1	−4.079
18.2	−6.526	23.2	−5.198	28.2	−4.058
18.3	−6.497	23.3	−5.173	28.3	−4.036
18.4	−6.468	23.4	−5.149	28.4	−4.015
18.5	−6.439	23.5	−5.125	28.5	−3.994
18.6	−6.410	23.6	−5.101	28.6	−3.972
18.7	−6.381	23.7	−5.077	28.7	−3.951
18.8	−6.353	23.8	−5.053	28.8	−3.930
18.9	−6.325	23.9	−5.029	28.9	−3.909
19.0	−6.296	24.0	−5.005	29.0	−3.888
19.1	−6.268	24.1	−4.981	29.1	−3.867
19.2	−6.240	24.2	−4.958	29.2	−3.846
19.3	−6.212	24.3	−4.934	29.3	−3.825
19.4	−6.184	24.4	−4.910	29.4	−3.804
19.5	−6.157	24.5	−4.887	29.5	−3.783
19.6	−6.129	24.6	−4.863	29.6	−3.762
19.7	−6.101	24.7	−4.840	29.7	−3.741
19.8	−6.074	24.8	−4.817	29.8	−3.720
19.9	−6.047	24.9	−4.793	29.9	−3.699

TABLE D-7   Omega Conversion Table (*Continued*)

p, %	db	p, %	db	p,%	db
30.0	−3.679	35.0	−2.687	40.0	−1.760
30.1	−3.658	35.1	−2.668	40.1	−1.742
30.2	−3.637	35.2	−2.649	40.2	−1.724
30.3	−3.617	35.3	−2.630	40.3	−1.706
30.4	−3.596	35.4	−2.611	40.4	−1.688
30.5	−3.576	35.5	−2.592	40.5	−1.670
30.6	−3.555	35.6	−2.573	40.6	−1.652
30.7	−3.535	35.7	−2.554	40.7	−1.634
30.8	−3.515	35.8	−2.536	40.8	−1.616
30.9	−3.494	35.9	−2.517	40.9	−1.598
31.0	−3.474	36.0	−2.498	41.0	−1.580
31.1	−3.454	36.1	−2.479	41.1	−1.562
31.2	−3.433	36.2	−2.460	41.2	−1.544
31.3	−3.413	36.3	−2.441	41.3	−1.526
31.4	−3.393	36.4	−2.423	41.4	−1.508
31.5	−3.373	36.5	−2.404	41.5	−1.490
31.6	−3.353	36.6	−2.385	41.6	−1.472
31.7	−3.333	36.7	−2.366	41.7	−1.454
31.8	−3.313	36.8	−2.348	41.8	−1.436
31.9	−3.293	36.9	−2.329	41.9	−1.419
32.0	−3.273	37.0	−2.310	42.0	−1.401
32.1	−3.253	37.1	−2.292	42.1	−1.383
32.2	−3.233	37.2	−2.273	42.2	−1.365
32.3	−3.213	37.3	−2.255	42.3	−1.347
32.4	−3.193	37.4	−2.236	42.4	−1.330
32.5	−3.173	37.5	−2.217	42.5	−1.312
32.6	−3.153	37.6	−2.199	42.6	−1.294
32.7	−3.134	37.7	−2.180	42.7	−1.276
32.8	−3.114	37.8	−2.162	42.8	−1.259
32.9	−3.094	37.9	−2.144	42.9	−1.241
33.0	−3.075	38.0	−2.125	43.0	−1.223
33.1	−3.055	38.1	−2.107	43.1	−1.205
33.2	−3.035	38.2	−2.088	43.2	−1.188
33.3	−3.016	38.3	−2.070	43.3	−1.170
33.4	−2.996	38.4	−2.051	43.4	−1.152
33.5	−2.977	38.5	−2.033	43.5	−1.135
33.6	−2.957	38.6	−2.015	43.6	−1.117
33.7	−2.938	38.7	−1.996	43.7	−1.009
33.8	−2.918	38.8	−1.978	43.8	−1.082
33.9	−2.899	38.9	−1.960	43.9	−1.064
34.0	−2.880	39.0	−1.942	44.0	−1.046
34.1	−2.860	39.1	−1.923	44.1	−1.029
34.2	−2.841	39.2	−1.905	44.2	−1.011
34.3	−2.822	39.3	−1.887	44.3	−0.994
34.4	−2.802	39.4	−1.869	44.4	−0.976
34.5	−2.783	39.5	−1.851	44.5	−0.958
34.6	−2.764	39.6	−1.832	44.6	−0.941
34.7	−2.745	39.7	−1.814	44.7	−0.923
34.8	−2.726	39.8	−1.796	44.8	−0.906
34.9	−2.707	39.9	−1.778	44.9	−0.888

TABLE D-7   Omega Conversion Table (*Continued*)

p, %	db	p, %	db	p,%	db
45.0	−0.871	50.0	0.000	55.0	0.872
45.1	−0.853	50.1	0.017	55.1	0.889
45.2	−0.835	50.2	0.035	55.2	0.907
45.3	−0.818	50.3	0.052	55.3	0.924
45.4	−0.800	50.4	0.069	55.4	0.942
45.5	−0.783	50.5	0.087	55.5	0.959
45.6	−0.765	50.6	0.104	55.6	0.977
45.7	−0.748	50.7	0.122	55.7	0.995
45.8	−0.730	50.8	0.139	55.8	1.012
45.9	−0.713	50.9	0.156	55.9	1.030
46.0	−0.695	51.0	0.174	56.0	1.047
46.1	−0.678	51.1	0.191	56.1	1.065
46.2	−0.660	51.2	0.209	56.2	1.083
46.3	−0.643	51.3	0.226	56.3	1.100
46.4	−0.625	51.4	0.243	56.4	1.118
46.5	−0.608	51.5	0.261	56.5	1.136
46.6	−0.591	51.6	0.278	56.6	1.153
46.7	−0.573	51.7	0.295	56.7	1.171
46.8	−0.556	51.8	0.313	56.8	1.189
46.9	−0.538	51.9	0.330	56.9	1.206
47.0	−0.521	52.0	0.348	57.0	1.224
47.1	−0.503	52.1	0.365	57.1	1.242
47.2	−0.486	52.2	0.382	57.2	1.260
47.3	−0.468	52.3	0.400	57.3	1.277
47.4	−0.451	52.4	0.417	57.4	1.295
47.5	−0.434	52.5	0.435	57.5	1.313
47.6	−0.416	52.6	0.452	57.6	1.331
47.7	−0.399	52.7	0.469	57.7	1.348
47.8	−0.381	52.8	0.487	57.8	1.366
47.9	−0.364	52.9	0.504	57.9	1.384
48.0	−0.347	53.0	0.522	58.0	1.402
48.1	−0.329	53.1	0.539	58.1	1.420
48.2	−0.312	53.2	0.557	58.2	1.437
48.3	−0.294	53.3	0.574	58.3	1.455
48.4	−0.277	53.4	0.592	58.4	1.473
48.5	−0.260	53.5	0.609	58.5	1.491
48.6	−0.242	53.6	0.626	58.6	1.509
48.7	−0.225	53.7	0.644	58.7	1.527
48.8	−0.208	53.8	0.661	58.8	1.545
48.9	−0.190	53.9	0.679	58.9	1.563
49.0	−0.173	54.0	0.696	59.0	1.581
49.1	−0.155	54.1	0.714	59.1	1.599
49.2	−0.138	54.2	0.731	59.2	1.617
49.3	−0.121	54.3	0.749	59.3	1.635
49.4	−0.103	54.4	0.679	59.4	1.653
49.5	−0.086	54.5	0.784	59.5	1.671
49.6	−0.068	54.6	0.801	59.6	1.689
49.7	−0.051	54.7	0.819	59.7	1.707
49.8	−0.034	54.8	0.836	59.8	1.725
49.9	−0.016	54.9	0.854	59.9	1.743

　　　　　　−Values　　　　+Values

TABLE D-7   Omega Conversion Table (*Continued*)

p, %	db	p, %	db	p,%	db
60.0	1.761	65.0	2.688	70.0	3.680
60.1	1.779	65.1	2.708	70.1	3.700
60.2	1.797	65.2	2.727	70.2	3.721
60.3	1.815	65.3	2.746	70.3	3.742
60.4	1.833	65.4	2.765	70.4	3.763
60.5	1.852	65.5	2.784	70.5	3.784
60.6	1.870	65.6	2.803	70.6	3.805
60.7	1.888	65.7	2.823	70.7	3.826
60.8	1.906	65.8	2.842	70.8	3.847
60.9	1.924	65.9	2.861	70.9	3.868
61.0	1.943	66.0	2.881	71.0	3.889
61.1	1.961	66.1	2.900	71.1	3.910
61.2	1.979	66.2	2.919	71.2	3.931
61.3	1.997	66.3	2.939	71.3	3.952
61.4	2.016	66.4	2.958	71.4	3.973
61.5	2.034	66.5	2.978	71.5	3.995
61.6	2.052	66.6	2.997	71.6	4.016
61.7	2.071	66.7	3.017	71.7	4.037
61.8	2.089	66.8	3.036	71.8	4.059
61.9	2.108	66.9	3.056	71.9	4.080
62.0	2.126	67.0	3.076	72.0	4.102
62.1	2.145	67.1	3.095	72.1	4.123
62.2	2.163	67.2	3.115	72.2	4.145
62.3	2.181	67.3	3.135	72.3	4.167
62.4	2.200	67.4	3.154	72.4	4.188
62.5	2.218	67.5	3.174	72.5	4.210
62.6	2.237	67.6	3.194	72.6	4.232
62.7	2.256	67.7	3.214	72.7	4.254
62.8	2.274	67.8	3.234	72.8	4.276
62.9	2.293	67.9	3.254	72.9	4.298
63.0	2.311	68.0	3.274	73.0	4.320
63.1	2.330	68.1	3.294	73.1	4.342
63.2	2.349	68.2	3.314	73.2	4.364
63.3	2.367	68.3	3.334	73.3	4.386
63.4	2.386	68.4	3.354	73.4	4.408
63.5	2.405	68.5	3.374	73.5	4.430
63.6	2.424	68.6	3.394	73.6	4.453
63.7	2.442	68.7	3.414	73.7	4.475
63.8	2.461	68.8	3.434	73.8	4.498
63.9	2.480	68.9	3.455	73.9	4.520
64.0	2.499	69.0	3.475	74.0	4.453
64.1	2.518	69.1	3.495	74.1	4.565
64.2	2.537	69.2	3.516	74.2	4.588
64.3	2.555	69.3	3.536	74.3	4.611
64.4	2.574	69.4	3.556	74.4	4.633
64.5	2.593	69.5	3.577	74.5	4.656
64.6	2.612	69.6	3.597	74.6	4.679
64.7	2.631	69.7	3.618	74.7	4.702
64.8	2.650	69.8	3.638	74.8	4.725
64.9	2.669	69.9	3.659	74.9	4.748

TABLE D-7   Omega Conversion Table (*Continued*)

p, %	db	p, %	db	p,%	db
75.0	4.771	80.0	6.021	85.0	7.533
75.1	4.794	80.1	6.048	85.1	7.567
75.2	4.818	80.2	6.075	85.2	7.602
75.3	4.841	80.3	6.102	85.3	7.686
75.4	4.864	80.4	6.130	85.4	7.671
75.5	4.888	80.5	6.158	85.5	7.706
75.6	4.911	80.6	6.185	85.6	7.741
75.7	4.935	80.7	6.213	85.7	7.776
75.8	4.959	80.8	6.241	85.8	7.812
75.9	4.982	80.9	6.269	85.9	7.848
76.0	5.006	81.0	6.297	86.0	7.884
76.1	5.030	81.1	6.326	86.1	7.920
76.2	5.054	81.2	6.354	86.2	7.956
76.3	5.078	81.3	6.382	86.3	7.993
76.4	5.102	81.4	6.411	86.4	8.080
76.5	5.126	81.5	6.440	86.5	8.067
76.6	5.150	81.6	6.469	86.6	8.104
76.7	5.174	81.7	6.498	86.7	8.142
76.8	5.199	81.8	6.527	86.8	8.179
76.9	5.223	81.9	6.556	86.9	8.217
77.0	5.248	82.0	6.585	87.0	8.256
77.1	5.272	82.1	6.615	87.1	8.294
77.2	5.297	82.2	6.645	87.2	8.333
77.3	5.322	82.3	6.674	87.3	8.372
77.4	5.346	82.4	6.704	87.4	8.411
77.5	5.371	82.5	6.734	87.5	8.451
77.6	5.396	82.6	6.764	87.6	8.491
77.7	5.421	82.7	6.795	87.7	8.531
77.8	5.446	82.8	6.825	87.8	8.571
77.9	5.471	82.9	6.856	87.9	8.612
78.0	5.497	83.0	6.886	88.0	8.653
78.1	5.522	83.1	6.917	88.1	8.694
78.2	5.548	83.2	6.948	88.2	8.786
78.3	5.573	83.3	6.979	88.3	8.778
78.4	5.599	83.4	7.011	88.4	8.820
78.5	5.624	83.5	7.042	88.5	8.862
78.6	5.650	83.6	7.074	88.6	8.905
78.7	5.676	83.7	7.105	88.7	8.948
78.8	5.702	83.8	7.137	88.8	8.992
78.9	5.728	83.9	7.169	88.9	9.036
79.0	5.754	84.0	7.202	89.0	9.080
79.1	5.780	84.1	7.234	89.1	9.125
79.2	5.807	84.2	7.267	89.2	9.169
79.3	5.833	84.3	7.299	89.3	9.215
79.4	5.860	84.4	7.332	89.4	9.260
79.5	5.886	84.5	7.365	89.5	9.306
79.6	5.913	84.6	7.398	89.6	9.353
79.7	5.940	84.7	7.432	89.7	9.400
79.8	5.967	84.8	7.466	89.8	9.447
79.9	5.994	84.9	7.499	89.9	9.494

TABLE D-7   Omega Conversion Table (*Continued*)

p, %	db	p, %	db	p,%	db
90.0	9.542	94.0	11.950	98.0	16.902
90.1	9.591	94.1	12.027	98.1	17.129
90.2	9.640	94.2	12.106	98.2	17.368
90.3	9.689	94.3	12.186	98.3	17.621
90.4	9.739	94.4	12.268	98.4	17.889
90.5	9.789	94.5	12.351	98.5	18.173
90.6	9.840	94.6	12.435	98.6	18.447
90.7	9.891	94.7	12.521	98.7	18.804
90.8	9.943	94.8	12.608	98.8	19.156
90.9	9.995	94.9	12.697	98.9	19.538
91.0	10.048	95.0	12.783	99.0	19.956
91.1	10.111	95.1	12.880	99.1	20.418
91.2	10.155	95.2	12.974	99.2	20.934
91.3	10.210	95.3	13.070	99.3	21.519
91.4	10.264	95.4	13.168	99.4	22.192
91.5	10.320	95.5	13.268	99.5	22.989
91.6	10.376	95.6	13.370	99.6	23.962
91.7	10.433	95.7	13.474	99.7	25.216
91.8	10.490	95.8	13.581	99.8	26.981
91.9	10.548	95.9	13.690	99.9	29.996
92.0	10.607	96.0	13.802	100.0	∞
92.1	10.666	96.1	13.917		
92.2	10.726	96.2	14.034		
92.3	10.787	96.3	14.154		
92.4	10.840	96.4	14.278		
92.5	10.911	96.5	14.405		
92.6	10.974	96.6	14.535		
92.7	11.038	96.7	14.669		
92.8	11.102	96.8	14.807		
92.9	11.168	96.9	14.950		
93.0	11.234	97.0	15.097		
93.1	11.301	97.1	15.248		
93.2	11.369	97.2	15.405		
93.3	11.438	97.3	15.567		
93.4	11.508	97.4	15.736		
93.5	11.579	97.5	15.911		
93.6	11.651	97.6	16.092		
93.7	11.724	97.7	16.282		
93.8	11.798	97.8	16.479		
93.9	11.873	97.9	16.686		

$$\Omega(\mathrm{db}) = 10 \log \left[\frac{p/100}{1-p/100} \right]$$

Proof 1 (see page 12)

$$S^2_{y;m} = S_{y;y}{}^2 + (\bar{y} - m)^2$$

variance of y with baseline of m = variance of y with baseline of $\bar{y}$ + square of the difference of average and m

$$\frac{\sum\limits_{i=1}^{n} (y_i - m)^2}{n} = \frac{\sum\limits_{i=1}^{n} (y_i - \bar{y})^2}{n} + (\bar{y} - m)^2$$

$$\Sigma (y_i - m)^2 = (y_i - \bar{y})^2 + n(\bar{y} - m)^2$$

$$\Sigma (y_i^2 - 2my_i + m^2) = \Sigma (y_i^2 - 2\bar{y}\, y_i + \bar{y}^2) + n(\bar{y}^2 - 2m\bar{y} + m^2)$$

$$\Sigma y_i^2 - 2m\, \Sigma y_i + nm^2 = \Sigma y_i^2 - 2\bar{y}\, \Sigma y_i + n\bar{y}^2 + n\bar{y}^2 - 2mn\bar{y} + nm^2$$

$$-2m\, \Sigma y_i = -2\bar{y}\, \Sigma y_i + 2n\bar{y}^2 - 2mn\bar{y}$$

$$-m\, \Sigma y_i = -\bar{y}\, \Sigma y_i + n\bar{y}^2 - mn\bar{y}$$

Given

$$\bar{y} = \frac{\Sigma y_i}{n}$$

$$-m\, \Sigma y_i = \frac{-\Sigma y_i\, (\Sigma y_i)}{n} + n \left(\frac{\Sigma y_i}{n}\right)^2 - mn \left(\frac{\Sigma y_i}{n}\right)$$

$$0 = -\frac{(\Sigma y_i)^2}{n} + \frac{(\Sigma y_i)^2}{n}$$

$$0 = 0$$

Proof 2 (see page 112)

$$SS_A = n(\overline{A}_1 - \overline{T})^2 + n(\overline{A}_2 - \overline{T})^2$$

Given: $n + n = N = 2n$

$$SS_A = n[(\overline{A}_1 - \overline{T})^2 + (\overline{A}_2 - \overline{T})^2]$$

$$= n([\overline{A}_1^2 - 2\overline{A}_1\overline{T} + \overline{T}^2 + \overline{A}_2^2 - 2\overline{A}_2\overline{T} + \overline{T}^2)$$

$$= n[\overline{A}_1^2 + \overline{A}_2^2 - 2\overline{T}(\overline{A}_1 + \overline{A}_2) + 2\overline{T}^2]$$

but $2\overline{T} = \overline{A}_1 + \overline{A}_2$

$$SS_A = n[\overline{A}_1^2 + \overline{A}_2^2 - 2\overline{T}(2\overline{T}) + 2\overline{T}^2]$$

$$= n(\overline{A}_1^2 + \overline{A}_2^2 - 4\overline{T}^2 + 2\overline{T}^2)$$

$$= n(\overline{A}_1^2 + \overline{A}_2^2 - 2\overline{T}^2)$$

$$= n\overline{A}_1^2 + n\overline{A}_2^2 - 2n\overline{T}^2$$

but $\overline{A}_1 = A_1/n; \overline{A}_2 = A_2/n;$ and $\overline{T} = T/2n$

$$SS_A = n\left(\frac{A_1}{n}\right)^2 + n\left(\frac{A_2}{n}\right)^2 - 2n\left(\frac{T}{2n}\right)^2$$

$$= \frac{A_1^2}{n} + \frac{A_2^2}{n} - \frac{T^2}{2n}$$

$$= \frac{A_1^2}{n} + \frac{A_2^2}{n} - \frac{T^2}{N}$$

This is the proof of the general formula for sum of squares using a two-level example.

but $T = A_1 + A_2$

$$SS_A = \frac{A_1^2}{n} + \frac{A_2^2}{n} - \frac{(A_1 + A_2)^2}{N}$$

$$= \frac{(A_1^2 + A_2^2)}{n} - \frac{(A_1^2 + 2A_1A_2 + A_2^2)}{2n}$$

$$= \frac{2(A_1^2 + A_2^2)}{2n} - \frac{(A_1^2 + 2A_1A_2 + A_2^2)}{2n}$$

$$= \frac{(2A_1^2 + 2A_2^2 - A_1^2 - 2A_1A_2 - A_2^2)}{2n}$$

$$= \frac{(A_1^2 + A_2^2 - 2A_1A_2)}{2n}$$

$$= \frac{(A_1^2 - 2A_1A_2 + A_2^2)}{N}$$

$$= \frac{(A_1 - A_2)^2}{N}$$

This is the proof of the specific formula for sum of squares.

Proof 3 (see page 117)

$$\sum_{i=1}^{N} (y_i - \overline{T})^2 = \left[\sum_{i=1}^{N} y_i^2 \right] - \frac{T^2}{N}$$

$$\Sigma (y_i^2 - 2\overline{T}y_i + \overline{T}^2) = \left[\sum_{i=1}^{N} y_i^2 \right] - \frac{T^2}{N}$$

$$\Sigma y_i^2 - \Sigma 2\overline{T}y_i + \Sigma \overline{T}^2 = \left[\sum_{i=1}^{N} y_i^2 \right] - \frac{T^2}{N}$$

$$\Sigma y_i^2 - 2\overline{T} \Sigma y_i + N\overline{T}^2 = \left[\sum_{i=1}^{N} y_i^2 \right] - \frac{T^2}{N}$$

Given $T = \Sigma y_i$ and $\overline{T} = T/N$

$$[\Sigma y_i^2] - 2\left(\frac{T}{N} \right) T + N\overline{T}^2 = \left[\sum_{i=1}^{N} y_i^2 \right] - \frac{T^2}{N}$$

$$[\Sigma y_i^2] - \frac{2T^2}{N} + \frac{T^2}{N} = \left[\sum_{i=1}^{N} y_i^2 \right] - \frac{T^2}{N}$$

$$[\Sigma y_i^2] - \frac{T^2}{N} = [\Sigma y_i^2] - \frac{T^2}{N}$$

Appendix

F

Definitions and Symbols

Definitions

Alpha (α) error   The probability that the null hypothesis will be rejected when it is in fact true; the chance of saying that a mean of a population is different when in fact it is equal to another population mean; type I error; the producer's risk of specifying a factor condition which is not really helpful.

Alternative hypothesis (H_a)   The assumption that the null hypothesis is not true; there is some inequality existing between the values being compared; the alternative hypothesis can be proven true at some confidence level which is always $(1-\alpha)$.

Analysis of variance (ANOVA)   A procedure to isolate one source of variation from another; a method to decompose the total variation present into accountable sources; a method to make a statistically based decision as to the causes of variation in an experiment.

Average   See *Mean of sample.*

Beta (ß) error   The probability that the null hypothesis will be accepted when it is in fact not true; the chance of saying there is no difference of two population means when in fact a difference does exist; type II error; the consumer's risk of a producer missing a helpful factor.

Binomial   The distribution of a discrete variable or attribute where there are two possible outcomes, the probability of which remains constant, such as tossing coins, passing or failing a test, fitting a go–no go gauge, having a defect present or not present, etc.

Blocking   A technique used in factorial experimentation to reduce the effect of unwanted variation by restricting the design of the experiment. A block of an experiment is a portion of the total experimental material that is expected to be more uniform than the whole because of more controlled conditions than if randomized. Treating a known source of variation as a factor.

Central limit theorem   Statistical theorem which states that sample means will tend to be normally distributed around the population mean with less variance than the individuals.

Component search   A procedure that is a derivative of factorial experiments in which components are interchanged between what is a good-performing unit and a poor-performing unit; in typical industrial situations only the poor unit is retested with other parts of unknown quality in an attempt to improve its performance.

Confidence interval (band)   The region enclosed by two limits of the performance characteristic which will contain the predicted mean at the stated confidence.

Confidence level   The value of one minus the risk involved $(1-\alpha)$ or $(1-\beta)$; stacking the odds in your favor that a correct decision will be made.

Confounded   To be unable to decompose certain sources of variation from one another in a particular experiment; a mixture of factorial effects within a particular column of an orthogonal array which are mathematically impossible to separate.

Continuous random variable   A variable which can randomly attain any value within some interval.

Control factor   A factor which can be set by a manufacturer and cannot be directly affected by the customer using the product or process.

Decomposition (of variance)   Taking the total amount of variation observed in an experiment and breaking it down into portions and assigning responsibility for that part of the variation.

Degrees of freedom (v)   The number of independent measurements available to estimate pieces of information; the number of independent (fair) comparisons that may be made within a set of data.

Design of experiments   See *Factorial experiment* and *Fractional-factorial experiment.*

Discrete random variable   A variable which can randomly attain only discrete values within some interval.

Experiment   An operation under controlled conditions to determine an unknown effect; to illustrate or verify a known law; a test to establish a hypothesis.

Experimental error   Residual error; the variation observed when products are tested under "identical" conditions, a portion of which is instrumentation (measurement) repeatability.

Factorial experiment   An experiment which extracts information about several design factors more efficiently than a traditional single-factor experiment.

Fractional-factorial experiment   A factorial experiment where only part of the treatment conditions are tested to more efficiently identify important factors; a Taguchi matrix is one example.

Histogram   A graphical representation of the sample frequency distribution where the width of the bar represents some interval of a variable and the height represents the frequency of occurrence of that value.

Inner noise Functionally related noise factors which affect variation within a product (wear, fade, hardening, etc.).

Interaction ($A \times B$) The synergistic effect of two or more factors in a factorial experiment. The effect of one factor depends on another factor.

Least square line A line fitted to a series of test observations such that the sum of the squares of the deviations of the test observations from the line is a minimum; the best-fitting line to test data.

Levels The settings of various factors in a factorial experiment; high and low values of pressure, temperature, etc.; may also have multiple levels of a factor; k equals the number of levels of a factor.

Linear graph A tool used in assigning factors and interactions to a Taguchi orthogonal matrix for the purpose of experimentation (see *Triangular tables*).

Loss function A continuous cost function which measures the cost impact of the variability of a product.

Main effect The effect of a factor acting on its own from one level to another to change the outcome in a factorial experiment.

Mean (population) (μ) The expected value of a population.

Mean (sample) ($\bar{x}, \bar{y}, \bar{z}$, etc.) The summation of test observations divided by the total number of observations; average.

Nested factor A factor which can be varied only over a portion of the entire experiment due to design or process limitations.

Noise factor A factor that disturbs the function of a product or process; factor which may be controlled during an experiment but in the customer's typical use may not be controlled by the manufacturer; a factor that the manufacturer wishes not to control.

Null hypothesis *(H_o)* Assumption that there is no inequality existing between the means of the values being compared.

Observation The result (data) obtained when an experiment is run at particular conditions; also response.

Orthogonal matrix (array) A fractional-factorial matrix which assures a balanced, fair comparison of levels of any factor or interaction of factors; all columns can be evaluated independently of one another.

Outer noise Environmentally related noise factors which affect variation within a product (temperature, humidity, operators, etc.).

Parameter design A method utilizing the appropriate level of a control factor (design parameter) to reduce the sensitivity to outer, inner, and product noises.

Polynomial decomposition Using equations to decompose the observed variation into the portions attributable to linear, quadratic, cubic, etc., functions of the factor being evaluated.

Pooling (degrees of freedom) In an analysis of variance, the combining of the sum of squares and the degrees of freedom of those factors and/or error estimates that are statistically insignificant to obtain a better estimate of experimental error.

Population   A group of similar items from which a sample is drawn for test purposes; a group of parts about which a decision is to be made.

Precision   The ability of a device to perform a measurement repeatedly with some defined variance.

Random variable   A variable which can assume any value from a set of possible values.

Repetition   Multiple test observations within a trial in an orthogonal matrix test.

Replication   Multiple test observations trial to trial in an orthogonal matrix test.

Residual error   See *Experimental error.*

Response   See *Observation.*

Sample   A random selection of items from a lot or population to evaluate the characteristics of that population; the sample must be representative of the population to be of any value.

Signal-to-noise ratio   A measure of the amount of observed variation present relative to the observed average of the data.

Standard deviation (population) (σ)   Square root of the variance of a population; measure of the scatter of the population around a mean.

Standard deviation (sample) (S)   Square root of the variance of a sample.

Sum of squares (SS)   The summation of the squared deviations relative to zero, to level means, or the grand mean of an experiment.

Tolerance design   To reduce variation of a performance characteristic by tightening tolerances on factors relating to performance; a method used when the parameter design efforts have proved to be insufficient in reducing variation.

Triangular tables   A tool used to locate the interacting columns in a Taguchi orthogonal matrix for the purpose of experimentation.

Variance (population) (σ^2)   Expected value of the square of the difference between the random variable and the mean of the population.

Variance (sample) (V or S^2)   Sum of the squares of the difference of each observation and the mean divided by the sample size minus one (for an unbiased estimate of variance); for a biased estimate of variance divide by sample size.

Symbols

A	factor of interest
A_1	sum of observations under condition A_1
$\overline{A}_1$	average of observations under condition A_1
α	alpha

β	beta
C	confidence level
F	random variable of F distribution; ratio of variances
H_a	alternative hypothesis
H_o	null hypothesis
HB	higher is better
k	the number of levels in factorial exp.; also the cost factor in the loss function
k_A	number of levels for factor A
L	loss per unit in the loss function
LB	lower is better
LSL	lower specification limit
m	nominal or target value of a characteristic
μ	population mean
ν	degrees of freedom
ν_A	degrees of freedom for factor A; equals $k_A - 1$
N	total number of observations or samples
n_{A_1}	sample size under conditions A_1
NB	nominal is best
σ	population standard deviation
σ^2	population variance
S	sample standard deviation
S^2	sample variance
SS	sum of squares
SS_T	sum of squares due to total variation
SS_m	sum of squares due to mean
SS_A	sum of squares due to factor A
SS_e	sum of squares due to error
S/N	signal-to-noise ratio
T	sum of all observations
$\overline{T}$	average of all observations
USL	upper specification limit
V	variance
W	class weight in accumulation analysis
y	individual observation or data point
y_i	ith observation
$\overline{y}$	sample mean; average observation

Example Experiments

This appendix contains several examples of both designing experiments and analyzing and interpreting actual results. The first example is a simple parameter design example; the next four examples are of designing experiments; and the last three examples are of design, analysis, and interpretation. As much as possible, the examples follow the experimental procedure outlined in Sec. 2-1-4. The experimental factors have been renamed to protect proprietary information.

Again, the emphasis is on the experimental design stage, since analysis and interpretation is very easy with a properly structured set of tests. How to test a product or process is generally known, and with the additional knowledge of a good experimental strategy, the results can be quite lucrative.

Example G-1   Engine Oil Fill Tube Cap

A contemporary engine oil fill tube cap, shown in Fig. G-1, must seal the tube against pressures generated in an engine crankcase yet be easy to install or remove. To effect a seal, the force to move the cap must be higher than the force created by crankcase pressure but be lower than a force that human fingers can provide for removal. The design of the cap provides installation and removal force by deflecting the rubber cap lip over the ridge inside the fill tube. The force required to remove or install is a function of the cap stiffness and the amount of deflection as shown by the graphs in Fig. G-2. The equation for the force is

$$\text{Force} = \text{constant} \times \text{deflection} \times \text{stiffness}$$

The constant is a function of the rubber properties and the stiffness a function of the design configuration.

The stiffness of the cap can be increased by increasing the wall thickness of the cap, and the deflection can be increased by increasing the difference between the OD and ID dimensions indicated in Fig. G-1. By proper selection of the aforementioned dimensions, two stiffness curves can be generated. A parameter design approach, however, would show a preference for the low

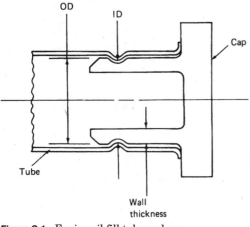

Figure G-1   Engine oil fill tube and cap.

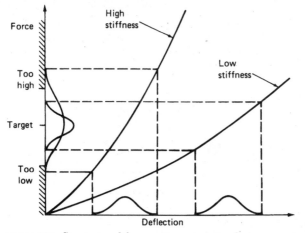

Figure G-2   Cap removal force using parameter design.

stiffness design because of the variation of the deflection from one assembly to another. The variance in the deflection is defined by the equation

$$\sigma^2_{\text{deflection}} = \sigma^2_{\text{OD}} + \sigma^2_{\text{ID}}$$

Variance is always additive in an assembly and is transmitted through to the variance in force by virtue of the stiffness relationship. The low-stiffness curve, with the same variation in deflection, transmits less variation in force and is, therefore, a higher-quality design, one that is more likely to meet force specifications. The loss function could be used to substantiate any cost penalties that might exist with the low-stiffness design, which would be a tolerance design approach.

Utilizing the lower-stiffness curve is similar to using the class II factors discussed in Sec. 7-5 of parameter design.

Example G-2   Windshield Washer Spray Nozzle

A windshield washer system must spray a pattern of solvent onto the glass that allows good cleaning of the windshield and is little affected by vehicle velocity. The vehicle velocity is a noise factor and the nozzle design, number of nozzle orifices, and position(s) of nozzles are control factors. The pattern tends to move (deflect) downward as velocity increases.

Problem. To improve the spray pattern and reduce the velocity effect on the windshield washer nozzles.

Objective. For a given vehicle design, develop a superior spray pattern and eliminate the velocity effect.

Measurement system. The spray pattern is classified by a numerical score (1 = poor, 2 = good, 3 = superior) and the deflection can be assessed by the difference in spray position when velocity increases from zero to 100 km/h (62 mph); a grid is marked on the inside of the windshield to serve as pattern size assessment and to determine deflection.

Factors and levels. The factor for position of the nozzles was broken into two factors: the fore-aft position and the side-to-side position. These two factors were to be evaluated over three levels to adequately cover the areas available for mounting nozzles (see Fig. G-3). The rearmost location is more

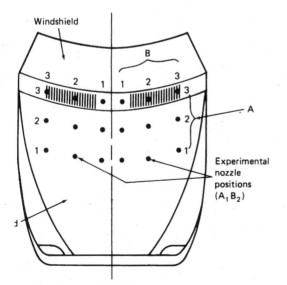

Figure G-3 Nozzle positions.

TABLE G-1   Windshield Washer Spray Factors and Levels

		Levels		
Factors		1	2	3
A Nozzle position (fore-aft)		1	2	3
B Nozzle position (side-to-side)		1	2	3
C Number of orifices		1	2	...
D Orifice diameter		.75 mm	1.00 mm	...

desirable from a complexity viewpoint, but the hood positions may be more desirable from a pattern or deflection viewpoint. Two other factors, number of nozzle orifices and orifice diameter, are evaluated only at two levels. Table G-1 shows the factors and levels.

Assignment to an array.   Since two of the factors are three-level and two others two-level, the total degrees of freedom is equal to 6 ($\nu_T = 6$) and should fit into an L9 experimental array. The two-level factors will have to be dummy-treated within this type of array as shown in Table G-2.

Conducting the experiment.   The trials are completely randomized since the nozzle is movable (magnetic mount) and the nozzle heads interchangeable. The tests are conducted by adjusting nozzle angles to hit a certain spot on the windshield at zero velocity. The quality of the pattern is then observed, the vehicle accelerated to 100 km/h, and the quality of the pattern observed once again along with the change in impact point. When the data is collected, the analysis of the pattern quality and deflection would take place separately.

TABLE G-2   Windshield Washer Spray Factor Assignment to an L9 Orthogonal Array

	Factors			
	A	B	C	D
	Column no.			
Trial no.	1	2	3	4
1	1	1	1	1
2	1	2	2	2
3	1	3	1′	2′
4	2	1	2	2′
5	2	2	1′	1
6	2	3	1	2
7	3	1	1′	2
8	3	2	1	2′
9	3	3	2	1

Example G-3   Connecting Rod Bolt Torque

Engine connecting rods are typically joined at the crankshaft end by bolts on either side of the bearing cap, as shown in Fig. G-4. Consistency of the retention torque is desirable to prevent loose bearing caps and subsequent spun bearings, which will cause catastrophic engine failure. Too high a nominal bolt torque could distort the bearing journal, which would be detrimental to bearing capacity, however.

Problem.   To develop a consistent means of initially obtaining and retaining a nominal bolt clamp load. To reduce distortion of the bearing journal under bolt loads.

Objective.   To develop a design nominal clamp load, 500 lb, with a consistent bolt torque, 70 lb•ft, and minimum distortion of the journal.

Measurement system.   The clamp load is estimated by measuring bolt stretch under load. The load versus deflection relationship must be known for the bolt design. The out-of-round condition is measured by taking several measurements either before and after torque is applied with an indicator gauge or on a special machine for roundness measurements.

Factor selection and levels.   Table G-3 summarizes the control and noise factors chosen in this experiment. The torque apply method, the number of times torque is applied, the initial torque, the thread lubrication, fit of the bolt, and seat surface finish were all selected as control factors and evaluated at two levels. Noise factors to the assembly process and bolt retention were nut surface finish and bolt hardness, which were also evaluated at two levels.

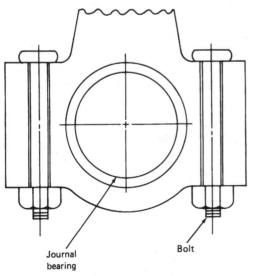

Journal
bearing

Bolt

Figure G-4   Connecting rod.

TABLE G-3   Connecting Rod Factors and Levels

Factors	Levels	
	1	2
A Torque apply method*	Air wrench	Hand wrench
B No. of times tightened*	3	1
C Initial torque*	100 N•m	80 N•m
D Thread lubrication*	Oiled	As supplied
E Bolt fit*	Loose	Tight
F Seat surface finish*	2 μm	5 μm
V Nut surface finish†	2 μm	5 μm
W Bolt hardness†	38 R_c	32 R_c

*Control factors requiring an L8 OA.
†Noise factors requiring an L4 OA.

In this situation, the levels were selected so that the first level should provide the most consistent load conditions; the engineers had wanted to specifically test a combination of treatment conditions. Since the first trial is all the first level of factors, the combination would be assured.

Assignment of factors.   The control factors were assigned to an L8 OA as the inner array and the noise factors assigned to an L4 OA as the outer array. Since one column is left over in the inner array, an interaction column was assigned for the experiment as shown in Table G-4.

Conducting the experiment.   Each combination of parts is assembled under the specified conditions for the trial and the amount of bolt stretch is determined, which is then translated into clamp load. Each connecting rod has two bolts. The out of roundness is determined for each connecting rod after torque is applied. The data arrangement is also shown in Table G-4.

Analyzing the data.   The raw data can be analyzed to determine which factors influence the average clamp load and out of roundness. The eight repetitions of load and four repetitions of out of roundness can be transformed to one signal-to-noise ratio for nominal is best and lower is better, respectively, and analyzed to determine which factors influence (reduce) variation.

Example G-4   Automobile Tire Experiment

Problem.   To reduce wear and improve traction of automobile tires on various automobiles.

Objective.   To reduce wear to less than 0.060 in after a tire wear test and to maximize the spin-out speed in a turning circle test.

Measurement system.   The amount of tread loss (after a certain amount of time on a duty cycle) determined with a depth micrometer and a fifth-wheel velocity indicated when spin-out occurs on the turning circle (the highest velocity obtained while still negotiating a specified turning radius).

TABLE G-4 Connecting Rod Factor Assignment

	A	B	A×B	C	D	E	F	V W	Load				Out of round				Signal-to-noise ratio	
									1 2 2 1				1 2 2 1				L	OOR
									1 2 1 2				1 2 1 2					
									1 1 2 2				1 1 2 2					
1	1	1	1	1	1	1	1		XX	XX	XX	XX	O	O	O	O	#	*
2	1	1	1	2	2	2	2		XX	XX	XX	XX	O	O	O	O	#	*
3	1	2	2	1	1	2	2		XX	XX	XX	XX	O	O	O	O	#	*
4	1	2	2	2	2	1	1		XX	XX	XX	XX	O	O	O	O	#	*
5	2	1	2	1	2	1	2		XX	XX	XX	XX	O	O	O	O	#	*
6	2	1	2	2	1	2	1		XX	XX	XX	XX	O	O	O	O	#	*
7	2	2	1	1	2	2	1		XX	XX	XX	XX	O	O	O	O	#	*
8	2	2	1	2	1	1	2		XX	XX	XX	XX	O	O	O	O	#	*

X = load data
O = out of roundness data
# = load S/N data
* = out-of-roundness S/N data

313

Factor selection. Two-level tire control factors *A, B, C, D, E,* and *F*; four-level factors *T* (type of tire), and *P* (position of tire on car).

Assignment of factors. The factors require a total of 12 degrees of freedom, 6 degrees of freedom for the two-level factors and 3 degrees of freedom for each of the four-level factors, so an L16 OA is required at a minimum. The required linear graph is shown in Fig. G-5 and the L16 standard linear graph in Fig. G-6. One option of merged columns is 1, 2, and 3 in addition to 7, 8, and 15. The experimental arrangement is shown in Table G-5.

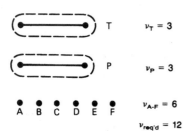

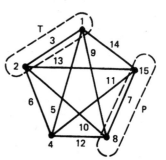

Figure G-5 Tire experiment required linear graph.

Figure G-6 L16 standard linear graph a.

TABLE G-5 **Tire Experiment Factor Assignment**

				Factors						
	T	*A*	*B*	*C*	*P*	*D*	*E*	*F*		
					Column no.					
Trial no.	1-2-3	4	5	6	7-8-15	9	10	11	12	13	14
1	1	1	1	1	1	1	1	1	1	1	1
2	1	1	1	1	2	2	2	2	2	2	2
3	1	2	2	2	3	1	1	1	2	2	2
4	1	2	2	2	4	2	2	2	1	1	1
5	2	1	1	2	3	1	2	2	1	1	2
6	2	1	1	2	4	2	1	1	2	2	1
7	2	2	2	1	1	1	2	2	2	2	1
8	2	2	2	1	2	2	1	1	1	1	2
9	3	1	2	1	3	2	1	2	1	2	1
10	3	1	2	1	4	1	2	1	2	1	2
11	3	2	1	2	1	2	1	2	2	1	2
12	3	2	1	2	2	1	2	1	1	2	1
13	4	1	2	2	1	2	2	1	1	2	2
14	4	1	2	2	2	1	1	2	2	1	1
15	4	2	1	1	3	2	2	1	2	1	1
16	4	2	1	1	4	1	1	2	1	2	2

Example G-5   Engine Piston Scuffing

Problem. Piston scuffing (abrasion, scoring) occurs during the final engine assembly test after initial engine build.

Objective. To eliminate or minimize piston scuffing.

Measurement system. The final assembly test is part of the measurement system and, because of the expense involved, will greatly restrain the number of tests that can be run. Two approaches can be used to assess scuffing:

1. An objective measurement such as percent scuffed area, engine torque loss, engine blow-by flow rate.

2. A subjective measurement such as severity rating (1 = very bad to 10 = very good, for instance), rank order of severity of scuffing and use rank number as data.

Factor selection and levels. This case was discussed at great length to review the possibilities for reducing the actual number of engine tests. The engine was a four-cylinder engine with sleeves, which allowed each cylinder to become a test site in itself, thus accomplishing four trials with one engine run. However, because of possible cylinder-to-cylinder variation, the experiment should be blocked on cylinders (treat the cylinder location as a four-level factor, factor K). To reduce the overall experimental time it was suggested to use two engines for the trials; one engine could be on the test stand while the other was being prepared for the next trial. Blocking again was recommended for the engines and treat engines as a two-level factor, factor L. The factors that were thought to influence scuffing were: A, piston/sleeve fit; B, piston ovality; C, piston contour; D, piston plating thickness; E, piston finish; F, piston material; G, piston cooling slot; H, sleeve finish; I, ring design; and J, engine rating. The engine rating, base and uprate, is a function of the head design and fuel control system, which are interchangeable on the basic block.

Assignment of factors. All of the factors plus the blocking items require 14 degrees of freedom, so an L16 OA is required. To make it easy to visualize the trial conditions, the engine rating was assigned to column 1 and columns 2, 4, and 6 were merged to form a four-level column for cylinder number. The remaining factors were assigned to the columns as shown in Table G-6.

Example G-6   Manual Transmission Lube System

A manual automotive transmission typically is lubricated by the rotating gears splashing oil throughout the entire mechanism. The oil level in the transmission is high enough to submerge the lower portion of some of the gears to accomplish splash lubrication. Other devices in the transmission may direct the oil to appropriate locations to provide uniform lubrication and cooling of components.

Problem. To evenly distribute the lubrication oil on critical components of a manual transmission.

TABLE G-6   Piston Scuffing Experiment Factor Assignment

						Factors						
	J	*K*	*A*	*B*	*C*	*D*	*E*	*F*	*G*	*H*	*I*	*L*
							Column no.					
Trial no.	1	2-4-6	3	5	7	8	9	10	11	12	13	14	15
1	1	1	1	1	1	1	1	1	1	1	1	1	1
2	1	1	1	1	1	2	2	2	2	2	2	2	2
3	1	2	1	2	2	1	1	1	1	2	2	2	2
4	1	2	1	2	2	2	2	2	2	1	1	1	1
5	1	3	2	1	2	1	1	2	2	1	1	2	2
6	1	3	2	1	2	2	2	1	1	2	2	1	1
7	1	4	2	2	1	1	1	2	2	2	2	1	1
8	1	4	2	2	1	2	2	1	1	1	1	2	2
9	2	1	2	2	2	1	2	1	2	1	2	1	2
10	2	1	2	2	2	2	1	2	1	2	1	2	1
11	2	2	2	1	1	1	2	1	2	2	1	2	1
12	2	2	2	1	1	2	1	2	1	1	2	1	2
13	2	3	1	2	1	1	2	2	1	1	2	2	1
14	2	3	1	2	1	2	1	1	2	2	1	1	2
15	2	4	1	1	2	1	2	2	1	2	1	1	2
16	2	4	1	1	2	2	1	1	2	1	2	2	1

Objective.   To achieve an average lubrication score of 25.0 or greater and to equally distribute the lubrication (no individual location below a value of 3.0).

Measurement system.   The transmission is operated in the various ranges at specific speeds and the amount of lubrication flowing to a location observed through plastic observation ports. The amount of lubrication flowing to a location is rated on a 0 to 5 scale (0 = none, 3.0 adequate, and 5 = abundant). The seven locations requiring lubrication are then totaled for an overall transmission score.

Factor selection and levels.   This experiment is set up with some noise factors included in the inner array, although this is not recommended. Four different gear ranges were selected with specific input speeds used in those ranges. A bearing location could use either a sealed or standard flow-through design and a trough for directing lubrication could be of an old or new design. The factors and levels are shown in Table G-7.

Assignment of factors.   A desire to investigate the $A \times B$, $A \times C$, and $A \times D$ interactions adds some interesting considerations to the column assignments. Figure G-7 shows the linear graphs required to evaluate the interactions when one of the factors is four levels. The modified L16 OA is provided in Table G-8 along with the assignment of factors and interactions.

Conducting the experiment.   The transmission specified by the trial condition is placed in the particular range for the test and the amount of lubrication observed at the seven critical sites. The total of the lubrication scores for the seven sites is also indicated in Table G-8. The individual scores were also

TABLE G-7   Factors for Manual Transmission Lube Experiment

Factors and levels
A_1 2d range at 3000 rpm
A_2 3d range at 3000 rpm
A_3 4th range at 2000 rpm
A_4 5th range at 2000 rpm
B_1 Sealed bearing
B_2 Standard bearing
C_1 No splash assist
C_2 Splash assist
D_1 New trough design
D_2 Old trough design

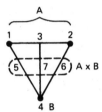

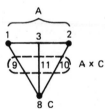

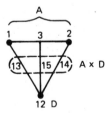

Figure G-7 Lubrication experiment linear graphs.

recorded and later analyzed, but for this example only the total score is addressed.

Analysis and interpretation. The analysis of variance for the lubrication score is summarized in Table G-9, and a pooled version of the ANOVA is summarized in Table G-10. The items that were pooled in the original table were all in the 6.00 or less range, while the two items that stand out as the largest were the use of the unsealed bearing design and the gear splash assist. These

TABLE G-8   Lube Experiment Factor Assignment

				Factors				
	A	B	A×B	C	A×C	D	A×D	
				Column no.				
Trial no.	1-2-3	4	5-6-7	8	9-10-11	12	13-14-15	Lube score
1	1	1	1 1 1	1	1 1 1	1	1 1 1	27.5
2	1	1	1 1 1	2	2 2 2	2	2 2 2	31.5
3	1	2	2 2 2	1	1 1 1	2	2 2 2	27.0
4	1	2	2 2 2	2	2 2 2	1	1 1 1	27.5
5	2	1	1 2 2	1	1 2 2	1	1 2 2	22.0
6	2	1	1 2 2	2	2 1 1	2	2 1 1	30.5
7	2	2	2 1 1	1	1 2 2	2	2 1 1	25.0
8	2	2	2 1 1	2	2 1 1	1	1 2 2	28.5
9	3	1	2 1 2	1	2 1 2	1	2 1 2	26.5
10	3	1	2 1 2	2	1 2 1	2	1 2 1	29.5
11	3	2	1 2 1	1	2 1 2	2	1 2 1	21.5
12	3	2	1 2 1	2	1 2 1	1	2 1 2	29.0
13	4	1	2 2 1	1	2 2 1	1	2 2 1	27.0
14	4	1	2 2 1	2	1 1 2	2	1 1 2	29.5
15	4	2	1 1 2	1	2 2 1	2	1 1 2	21.0
16	4	2	1 1 2	2	1 1 2	1	2 2 1	25.0

TABLE G-9   ANOVA Summary for
Lubrication Score

Source	SS	ν	V
A	15.92	3	5.31
B	23.77	1	23.77
C	70.14	1	70.14
D	0.39	1	0.39
A×B	16.68	3	5.56
A×C	9.06	3	3.02
A×D	14.55	3	4.85
Total	150.48	15	

TABLE G-10   Pooled ANOVA Summary for
Lubrication Score

Source	SS	ν	V	F	P
B	23.77	1	23.77	5.46‡	12.90
C	70.14	1	70.14	16.12#	43.72
e_p	56.58	13	4.35		43.38
Total	150.48	15			100.00

†At least 90% confidence.
‡At least 95% confidence.
#At least 99% confidence.

two items contributed over half of the variation in the observed results. The estimate for the average of the B_1C_2 condition is:

$$\hat{\mu}_{B_1C_2} = \overline{B}_1 + \overline{C}_2 - \overline{T}$$
$$= 28.00 + 28.88 - 26.78 = 30.10$$

The confidence interval for this condition is

$$CI = \sqrt{(F_{.10;1;13} \times V_e)/n_{eff}}$$
$$= \sqrt{(3.14 \times 4.35)/(16/3)} = 1.60$$

The average lubrication score is then estimated to fall somewhere in the range of 28.5 to 31.70 with 90% confidence. The objective of the experiment was to achieve an average of at least 25.0, so that objective potentially can be met with those two design features.

Confirmation experiment. The confirmation experiment is necessary to verify the conclusions of the screening experiment. Several transmissions should be assembled with the unsealed bearing and splash assist and tested to prove the theory.

Example G-7   Spot Welding Development

During the process of spot welding, pieces of sheet metal are joined together by forcing a high current through a localized area of contact; the thin metal is often distorted for various reasons. Often parts are held in jigs for locating purposes, but the heating and pressures used to hold parts still result in a warped workpiece. A research and development proposal of performing the spot welding through a sealant material was thought to make the distortion potentially worse.

Problem. Distortion of part during the spot welding process.

Objective. To minimize the distortion of experimental sheet metal components during spot welding to a value of less than 0.75 mm on the average and 1.00 mm for any individual value.

Measurement system. A height gauge is used to measure the right and left ends of components held in a fixture prior to and after welding to obtain welding distortion. The welding fixture is shown in Fig. G-8.

Factor selection and levels. The factors and levels selected for the welding experiment are shown in Table G-11. The weld schedule includes two different arrangements of welding contacts, welding pressure, and amperage.

Assignment of factors. Since there are 7 two-level factors, the experiment can use an L8 OA with all columns assigned.

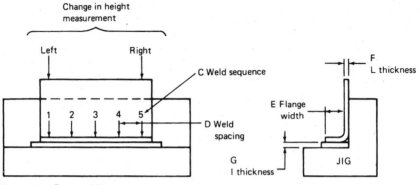

Figure G-8 Spot welding experiment.

TABLE G-11   Spot Welding Factors and Levels

	Factors	Levels 1	2
A	Weld schedule	1	2
B	Sealant thickness	3 mm	5 mm
C	Weld sequence	1-2-3-4-5	1-5-2-4-3
D	Weld spacing	20 mm	30 mm
E	Flange width	10 mm	15 mm
F	L thickness	0.75 mm	1.00 mm
G	I thickness	1.00 mm	1.5 mm
N	Measurement position (noise)	Left	Right

Conducting the experiment. The experiment was accomplished in two blocks, since changing the welding schedule was somewhat difficult. The remaining items are very simple to change from trial to trial and were completely randomized within the blocks. Five parts were manufactured under the trial conditions, and the change in height measured at both the right and left ends before and after welding. The resultant information is summarized in Table G-12, and the data is coded by multiplying each of the deflection values by 100 to allow computer analysis.

Analysis and interpretation. A series of ANOVA summary information is shown in Tables G-13 to G-17, which include the unpooled and pooled versions of the ANOVA of raw data and S/N data, in that order. The analysis of the raw data indicates that up to five of the seven factors contribute something to the variation in deflection; however, two of the factors make a very small contribution as shown in Table G-14. Further pooling of the column effects as shown in Table G-15 indicates that only three factors are really contributing the majority of the variation in deflection. The first weld schedule, the thinner sealant thickness, and the thicker I component reduce the average deflection

TABLE G-12 Spot Welding Experimental Data

Trial no.	Left end					Right end					S/N
1	100*	97	76	87	80	82	71	59	68	58	−37.95
2	61	66	62	60	73	51	56	56	52	56	−35.51
3	79	72	74	65	67	81	82	78	78	74	−37.52
4	96	102	116	111	108	103	106	109	124	99	−40.64
5	75	89	73	79	67	75	88	68	71	67	−37.57
6	95	102	89	88	84	97	100	85	94	80	−39.24
7	109	115	107	109	110	113	111	94	106	106	−40.68
8	98	125	90	109	108	74	94	70	103	110	−39.95

*Deflection in mm × 100.

TABLE G-13 ANOVA Summary for Spot Welding Experiment

Source	SS	ν	V	F	P
A	3537.81	1	3537.81	45.66	12.18
B	8988.81	1	8988.81	116.02	31.37
C	42.06	1	42.06	0.54	
D	510.06	1	510.06	6.58	1.52
E	72.19	1	72.19	0.93	
F	768.81	1	768.81	9.92	2.43
G	7411.25	1	7411.25	95.66	25.81
T_1	21,330.99	7			
N	627.19	1	627.19	8.10	1.93
$A{\times}N$	0.44	1	0.44	0.01	
$B{\times}N$	162.44	1	162.44	2.10	0.30
$C{\times}N$	980.00	1	980.00	12.65	3.18
$D{\times}N$	7.19	1	7.19	0.09	
$E{\times}N$	61.25	1	61.25	0.79	
$F{\times}N$	281.25	1	281.25	3.63	0.72
$G{\times}N$	0.81	1	0.81	0.01	
T_2	23,451.56	15			
e	4958.38	64	77.47		
T_3	28,409.94	79			

TABLE G-14 Partially Pooled ANOVA Summary for Spot Welding Experiment

Source	SS	ν	V	F	P
A	3537.81	1	3537.81	45.60#	12.18
B	8988.81	1	8988.81	115.86#	31.37
D	510.06	1	510.06	6.57‡	1.52
F	768.81	1	768.81	9.91#	2.43
G	7411.25	1	7411.25	95.53#	25.81
N	627.19	1	627.19	8.03#	1.93
$C{\times}N$	980.00	1	980.00	12.63#	3.18
e_p	5586.01	72	77.58		21.58
T	28,409.94	79			100.00

†At least 90% confidence.
‡At least 95% confidence.
#At least 99% confidence.

TABLE G-15   Completely Pooled ANOVA Summary for Spot Welding Experiment

Source	SS	ν	V	F	P
A	3537.81	1	3537.81	31.74#	12.06
B	8988.81	1	8988.81	80.64#	31.25
G	7411.25	1	7411.25	66.48#	25.69
e_p	8472.07	76	111.47		31.00
T	28,409.94	79			100.00

†At least 90% confidence.
‡At least 95% confidence.
#At least 99% confidence.

TABLE G-16   S/N ANOVA for Spot Welding Experiment

Source	SS	ν	V
A	4.22	1	4.22
B	9.07	1	9.07
C	0.10	1	0.10
D	0.33	1	0.33
E	0.01	1	0.01
F	1.24	1	1.24
G	7.93	1	7.93
T	22.90	7	

TABLE G-17   Pooled S/N ANOVA for Spot Welding Experiment

Source	SS	ν	V	F	P
A	4.22	1	4.22	10.11‡	16.60
B	9.07	1	9.07	21.73#	37.78
G	7.93	1	7.93	19.00‡	32.81
e_p	1.68	4	0.42		12.81
T	22.90	7			100.00

†At least 90% confidence.
‡At least 95% confidence.
#At least 99% confidence.

during the welding process since deflection is a lower-is-better characteristic. The ANOVA tables indicate which columns are important, but the level averages indicate which level is superior. The estimate of the mean is

$$\hat{\mu}_{A_1 B_1 G_2} = \overline{A}_1 + \overline{B}_1 + \overline{G}_2 - 2\overline{T}$$
$$= 79.88 + 75.93 + 76.90 - 2(86.53) = 59.65$$

The confidence interval for this condition is

$$CI = \sqrt{(F_{.10;1;76} \times V_e)/n_{\text{eff}}}$$

$$= \sqrt{(2.78 \times 111.47)/(80/4)} = 3.94$$

The average with 90% confidence falls within the range of 0.5571 to 0.6359 mm when the data is uncoded by dividing by 100. The variance of 111.47 provides a standard deviation of 10.56 mm (0.1056 mm when uncoded). If the average is as high as 0.6359 and three standard deviations are added to that value, the predicted value is 0.9527 mm, which is within the objective of the experiment. The average value and individual values are predicted to meet the experimental objectives.

An interpretation of the S/N data indicates that the same factors and levels increase the S/N value as shown in Tables G-16 and G-17. The average S/N value predicted for $A_1B_1G_2$ is -35.85 db. This is a situation typically encountered with either the S/N ratio for LB or HB characteristics; the factors which decrease or increase averages, respectively, will increase the appropriate S/N ratio. The analysis of S/N data, therefore, is not as meaningful as in the case of NB characteristics.

For the condition $A_1B_1G_2$, one can see that trial 2 has that treatment condition and the average results for only five repetitions fall within the confidence interval for the average of both the raw data and S/N data. Trial 2 is like a confirmation experiment within the original experiment; the raw data varies from 0.51 to 0.73 mm deflection, which gives great promise for the final confirmation experiment that may be run with a larger number of parts.

Example G-8   Brake Material Bonding Experiment

Brake friction material is chemically bonded to a steel brake shoe to provide support for the friction material during operation. The integrity of the bonding process is critical to the function of the brake shoe assembly. The process fundamentally consists of cleaning, applying adhesive, and drying the individual parts, which are then joined in an assembly and heated to activate the chemical bonding. At the end of the bonding process, samples of the batch of assemblies are tested for shear strength of the bond in a special test fixture. An experiment was proposed to increase the bond strength as much as possible (HB).

Problem.   Insufficient bond strength on the final assembly bond tester.

Objective.   To achieve a minimum bond shear strength of 150 lb.

Measurement system.   A section of the brake shoe assembly is cut off, mounted in the text fixture, and load applied until shear failure is induced.

Factor selection and levels.   The process consists of many steps; subsequently, several factors are possible for evaluation. The 15 factors and levels chosen for the experiment are shown in Table G-18. In this situation, some of the levels of the factors cause the chosen factor to actually become a noise factor. For

TABLE G-18   Brake Shoe Bonding Factors and Levels

		Levels	
Factors		1	2
Shoe			
K	Shot blast	1 min	3.5 min
A	Solvent wash	Butyl alcohol	MEK
B	Drying time	15 min	30 min
C	Shoe adhesive	Old	New
D	Drying time	2 h	4 h
Lining			
E	Solvent wash	Butyl alcohol	MEK
F	Drying time	30 min	60 min
G	Lining sealer	Old	New
H	Drying time	2 h	4 h
I	Lining adhesive	Old	New
J	Drying time	2 h	4 h
Brake shoe assembly			
L	Clamp load	500 lb	600 lb
M	Curing time	30 min	120 min
O	Curing temperature	360°F	400°F
N	Cooling time	30 min	60 min

instance, the age of the adhesives is really a noise to the bonding process. If the age turns out to be statistically significant and contributes a large percentage to variation, then the age of the adhesives will have controlled more closely.

Assignment of factors. The factors are going to completely saturate an L16 OA, but in this case the assignment was not in an alphabetical order. The factors were assigned sequentially to columns 1 to 15 in this order: *D, J, O, B, C, A, F, G, E, I, K, L, M, N,* and *H.* Since these were all two-level factors, the OA will not be shown.

Conducting the experiment. The brake shoe assemblies were made according to the trial "recipes" and shear-tested in the appropriate fixture. The load results for three different shoes in a top and bottom position are shown in Table G-19. In this situation, position is treated as a noise factor *P.*

Analysis and interpretation. Two points are very clear from a cursory analysis of the data. One, there is a lot of variation within a trial; two, half of the trials registered zero or very nearly zero loads, which means one factor is substantially affecting the results. Concerning the second point, one of the levels of a factor could be outside the operating window for that factor. For this reason, care must be taken not to select levels too widely apart. A pooled ANOVA summary is shown in Table G-20. The one factor that controls nearly half of the observed variation is *M,* the final curing time, and the second level provides substantially superior loads relative to the first level. The 30-min cure time is obviously inadequate, but some other cure time between 30 and

TABLE G-19   Brake Shoe Assembly Shear Force
Results

Trial no.	Top			Bottom		
1	0*	0	0	0	0	0
2	40	74	52	12	57	15
3	23	8	14	61	9	70
4	0	0	0	0	0	0
5	0	0	0	0	0	0
6	106	122	68	92	131	47
7	9	96	64	11	32	11
8	0	8	9	0	3	3
9	124	165	69	82	52	11
10	0	6	6	0	3	4
11	0	6	4	0	6	5
12	81	73	51	67	8	21
13	9	123	21	103	6	22
14	0	0	0	0	0	0
15	0	1	0	0	0	0
16	7	11	26	62	68	65

*Shear force, lbs.

TABLE G-20   Pooled ANOVA Summary for Brake Shoe Experiment

Source	SS	ν	V	F	P
O	5148.02	1	5148.02	9.13#	3.22
B	4043.01	1	4043.01	7.17#	2.44
F	1708.59	1	1708.59	3.03†	0.80
E	4746.09	1	4746.09	8.41#	2.94
I	2137.59	1	2137.59	3.79†	1.11
M	64,740.09	1	64,740.09	114.78#	45.10
N	3492.09	1	3492.09	6.19‡	2.06
$O{\times}P$	4069.00	1	4069.00	7.21#	2.46
$N{\times}P$	3712.59	1	3712.59	6.58‡	2.21
e_p	48,507.18	86	564.04		37.66
T	142,304.25	95			100.00

†At least 90% confidence.
‡At least 95% confidence.
#At least 99% confidence.

120 min may increase shear strength suddenly. The average shear strength under the 120-min cure condition was only 53.27 lb, which means the experimental objective cannot be met with only one factor unless an increased cure time beyond 120 min causes a dramatic rise in shear strength.

Another note on the ANOVA results: The error variance of the experiment is quite large. The standard deviation is 23.75 lb, which provides a six-sigma span of 142.5 lb for the measurement discrimination. A recommended follow-up from this experiment would be to investigate the repeatability of the shear test machine to reduce the experimental error as much as possible. The variability of the measuring device could be masking the presence of some other contributing factors as indicated by the low percentage contributions; recall that error variance effectively subtracts from the factor sum of squares. If the measuring device has good repeatability, a high S/N ratio, then there are other factors not included in the original screening experiment that are contributing to the observed variation. In addition, the repeatability of the shear tester may not be the only problem with it. The calibration should also be checked, since that may account for the low loads obtained.

Summary. This example was included to demonstrate the kinds of problems encountered in experimentation. Not all experiments succeed, but the two most common reasons are the lack of repeatability of the measuring system(s) used in obtaining the results and the exclusion of influential factors due to being unaware or choosing to minimize the size of the experiment by reducing the number of factors.

Index

ABOUT THE AUTHOR

Phillip J. Ross is a mechanical engineering graduate of General Motors Institute (BME 1970). The majority of his career has been associated with General Motors automotive powertrain industry from the product design/development aspect. First working with Allison Transmission Division from 1970 to 1987 and then with Saturn Corporation until the present time. Assignments have included working in the design phase of many transmission components and systems (he is the holder of three product design patents), developing statistical/quality methods and training, and performing process development including lost foam casting, painting, molding, and others.

Mr. Ross is a Certified Quality Engineer (ASQC) and has managed a consulting profession since 1985. He has taught statistical and quality classes in the United States, Great Britain, Holland, Japan, and Singapore accumulating over 2000 hours of time in classroom training. Clients have varied over several industries including Raymark (friction materials), Reynolds Aluminum, Steel Founder's Society of America, Georgetown Steel, Grede foundries (gray iron), ECC International (kaolin clay), Oxford Instruments (superconducting wire), Bohn Aluminum (die castings), Cordis (heart catheters), and others.

Mr. Ross has also authored articles for professional journals such as "The Role of Taguchi Methods and Design of Experiments in QFD," which was published in June 1988 in *Quality Progress* by ASQC, and "Education, Training, and Implementation of Design of Experiments," which was published in the Fall 1988 issue of *Target* by AME.